WOE TO THE UNWARY

WOE
TO THE
UNWARY

A Memoir of
Low Level Bombing Operations 1941

Roy Conyers Nesbit

WILLIAM KIMBER · LONDON

First published in 1981 by
WILLIAM KIMBER & CO. LIMITED
Godolphin House, 22a Queen Anne's Gate,
London, SW1H 9AE

© Roy Conyers Nesbit, 1981
ISBN 0-7183-0348-2

Photoset by Robcroft Ltd, London SW1
and printed in Great Britain by
Redwood Burn Limited, Trowbridge

Contents

List of Illustrations

FOR

Squadron Leader John Frederick Percival

'Do not let us speak of darker days; let us rather speak of sterner days. These are not dark days, these are great days – the greatest days our country has ever lived, and we must all thank God that we have been allowed, each according to our stations, to play a part in making these days memorable in the history of our race.'

WINSTON CHURCHILL, *29th October 1941*

Acknowledgments

In researching the background material for this book, I have become indebted to Mike Tagg of the RAF Museum, Ted Hine of the Photographic Department of the Imperial War Museum, and Squadron Leader F.W. Joe Davies of the Air Historical Branch at the Ministry of Defence, all of whom have contributed extensive specialist knowledge willingly and cheerfully. British Aerospace, Aircraft Group, Weybridge-Bristol Division, have also been most helpful in providing their advice and many indispensable photographs.

I am deeply grateful for the help given to me by fellow-survivors of 217 Squadron, Wing Commander A.H. 'Junior' Simmonds, DFC, Flight Lieutenant G. Alan Etheridge, DFC, Flight Lieutenant Stanley Clayton, LLB, and Wing Commander Arthur Aldridge, DFC and bar, MA.

Another long-standing friend from RAF days, Flight Lieutenant J. Dudley F. Cowderoy, BSc. (Eng), FICE, AIAS, MSocE. (France) helped me considerably with technical matters, whilst his wife Stella encouraged me to embark on this book.

My brother John K. Nesbit, FI Struct E, AIAS, MSoE (France) MConsE((France) helped me to collate the information, whilst his wife Marjorie patiently typed the manuscript. Their son, Peter Nesbit BA, designed one book jacket.

Lastly, I am more than grateful to Bert and Margaret Angell for allowing me to convert a room in their South Kensington flat into a miniature Operations Room whilst this book was being written.

Any errors which remain in this book after this considerable help are my responsibility alone.

R.N.

South Kensington.
September 1980.

Author's Note

When my mother died in 1979 she was the last of a long-lived generation in our family. During the sad task of clearing up her affairs, my brothers and I sifted through many items of memorabilia and decided that we would write a record before everything was irretrievably dispersed. This led us to genealogical research and the construction of a family history, a process which intensifies in fascination and may take many more years of study.

The last of these documents was a batch concerning myself. There was an old suitcase containing innumerable photographs taken during World War Two, my flying log book of the period, and most of my navigation logs preserved without official permission from my operational work on Bristol Beauforts in 1941. I had rarely discussed this year of action with anyone apart from ex-RAF friends when we would occasionally exchange wartime recollections, but after some hesitation I decided to write a short account of what had been an extraordinary period in my life. Although reluctant to inflict my reminiscences on friends and relatives I knew that I would be intensely interested if I could read, say, an account of my forebears' experiences at the Battle of Flodden in 1531, near where they lived on the Scottish border.

The navigation logs are not well-written. It was difficult to hold pencil to paper in the turbulence and vibration of a Beaufort flying at low level, especially when gripped in a hand encased in maybe three layers of gloves, under a tiny light over a chart table in an otherwise darkened aircraft, and there were times in action when the attempt to write had to be abandoned altogether. But re-reading these old logs and re-plotting the courses we flew, many half-forgotten episodes flooded back into my mind and I could remember with what seemed like complete clarity the numerous but brief moments of action in which I was involved. The exhilaration, comradeship, fear and tiredness that were part of our daily life could be re-captured vividly as I traced the courses of flight after flight, and remembered the faces of those who lived and died in the squadron in that year.

Biographies of the last war almost invariably concern those who made some notable or dramatic contribution to its result. I am not to be included in such company. If I have any merit at all, it is as a recorder of events and as a survivor, and certainly not as an unsung hero. If any remarkable character emerges from these pages it is that of my Canadian pilot, John Percival, although he would have been the last to cast himself in such a role.

My navigation logs and my flying log book have been checked with 217 Squadron Operational Record Book at the Public Record Office at Kew. It is interesting for me to find that this Operational Record Book is wrong in several respects. Five of my operational sorties are not recorded at all. On two other operational flights another navigator is incorrectly shown in my place. I am shown as having navigated on an operational flight on which I definately did not fly. Several incidents are recorded in the wrong sequence or are attributed to the wrong aircraft, and some episodes which might show the squadron in an unfavourable light have been conveniently left out. Many of these errors were made when our squadron adjutant had temporarily passed over his duties to someone else, but also distortions probably occurred because the information was entered third-hand. Crews were de-briefed in the Operational Room, and the information was collected by the squadron intelligence officer and passed to the squadron adjutant, who was trying to record confusing events in a period when the squadron was employed on operational work at all hours of day and night and when our aerodrome was subjected to frequent bombing attacks. In spite of these errors and omissions the Operational Record Book is a startling document and I found it fascinating.

There are no fictitious names in this narrative. Where the naming of certain people might cause distress, I have left those concerned unnamed and have been imprecise as to dates. Otherwise, all of the episodes are accurately related, or so I believe, but if there are any faults they must be ascribed to tricks in my memory rather than my intention.

CHAPTER ONE

Involuntary Navigator

'Youth is a blunder, manhood a struggle, old age a regret.'
Benjamin Disraeli (1844)

In 1939 the prospect of war with Germany held many attractions for young men in the United Kingdom and the Commonwealth. Some saw the forthcoming conflict as an opportunity to escape from the dull routine of their junior positions at work, whilst others looked forward to the end of their schooldays and longed for acceptance into some glamorous branch of the services where their smart uniforms would be the envy of their companions and the pride of their parents. Moreover, patriotism was mixed with this self-interest, for it was easy to regard the Nazi philosophy with abhorrence and the pacifism of the 1930's with contempt. It was perhaps fortunate for the United Kingdom that a flood of rather naive but not unintelligent young men volunteered for the most dangerous tasks that the country could offer and that this patriotic surge continued throughout the war.

Of all the branches of the services the Royal Air Force probably attracted most of such recruits. At the age of eighteen I had already applied for the Volunteer Reserve and through this means hoped to obtain a short-service commission. As an assiduous reader of the exploits of the Great War flying aces and a keen watcher of the new Hurricanes and Spitfires at North Weald aerodrome, near where my parents lived, I saw myself as a fighter pilot possessed of such skill and daring that both the· Luftwaffe and my country would be astounded at the ease with which I bagged enemy bomber after bomber, not to mention their clumsy fighter escorts. My dismal work as a junior clerk in Lloyds Bank in Fenchurch Street, where I was employed preparatory to joining my father in the Bank of England, would fade into the background and I would bask in the adulation of a grateful country. It was therefore with happy enthusiasm that I listened to the words of Neville Chamberlain's broadcast at 11 a.m. on 3rd September 1939, and in company with an old schoolfriend cycled all the way to Romford where we knew that a recruiting office was open on Sunday.

The following Tuesday, I received with a mixture of delight and slight apprehension, a request to present myself at RAF Station Uxbridge for a medical examination. My physique was not impressive, being a rather insignificant and skinny 5′8″ but I consoled myself with the firm knowledge that many Great War pilots had been small men. Also, I had secretly purchased from HM Stationery Office a copy of Air Publication 130, *The Medical Examination for Fitness for Flying*, and had practised for months some of the specified tests, developing the ability to hold my breath for at least ninety seconds as well as achieving forty press-ups without stopping. Only one disability worried me, for, extraordinary as it may seem these days, it was considered very unfortunate indeed to be born left-handed. As a schoolboy I had been forced to change from writing with my left hand to my right hand, and was not even permitted to bat left-handed at cricket. As a consequence, my classroom and sports performance deteriorated alarmingly and for a while I developed a marked squint which was gradually corrected with the aid of temporary spectacles but which left me with a smarting resentment of schoolmasters and sportsmasters and which still rankles with me. But in the event, this handicap remained concealed during the medical examination and was only detected months later when, to the fury of a drill sergeant, I attempted to fire a rifle left-handed instead of applying the butt to my right shoulder in the conventional and acceptable manner.

Stiff though the medical requirements were, I passed them without difficulty whilst many more robust young men were rejected, some of them even bursting into tears when told that they were slightly colourblind or had some minor physical defect. And so I was duly sworn in as an Aircraftman Second Class, enlisted as Aircrew (Under Training) and directed to a hut named Ypres to spend the night preparatory to kitting out. There was little to do except sample a rather unsavoury meal and to listen to the obscene but interesting songs bawled by the airmen in neighbouring huts and to comment with fellow recruits on what we regarded as the moronic attitude of the corporals in charge of us. The following day, I received a series of injections and was handed an ancient second-hand uniform, to my disappointment, and was further mortified to be told to go home on paid leave and await further orders. On my arrival at home, my mother opened the door and to my surprise registered distress instead of pleasure at the sight of me in uniform. It took me a few minutes to realise that the memory of her two brothers mutilated on the Western Front, my father's brother killed at Second Ypres, my elder brother already called up from the Territorials, and my twin younger brothers approaching military age, might have something to do with her foreboding. I was just eighteen and looked a lot younger, even in uniform.

For fourteen long weeks I fretted at home, reading anything that concerned flying training but bewildered by the 'phoney war', that strange period of inactivity between the Anglo-French and German fronts that persisted until the Blitzkrieg was launched the following April. Finally, I was instructed to report to Uxbridge on New Year's Day 1940, and went off enthusiastically to war in my ancient uniform, carrying an Airman's Diary for 1940 and a 'Stop-a-Shot' steel mirror in my left breast pocket, both of these articles having been presented to me by my father, who was a veteran of the Great War.

At Uxbridge I joined a group of cadets and entrained for Cambridge, where I found myself a member of 11 Flight, C Squadron, Downing College. This was part of the university, half of the college still being occupied by undergraduates reading theology. It was a large four-sided building surrounding a quadrangle, which served as our drill ground. Our quarters overlooked this quadrangle and were unheated throughout that intensely cold winter and, in the spartan RAF tradition, we were supposed to keep the windows open to reduce epidemics, whilst we slept wrapped in newspapers and sodden blankets.

On the first day after our arrival, we were introduced to our drill corporal, a compact, nutty Cockney with a steely eye. He looked at us with disfavour and disbelief.

'My Gawd', were his first words. 'What 'ave we got 'ere? U/T pilots? I don't believe it! 'Ow we goin' to win this bleedin' war?'

Then he straightened up, snapped to attention and roared:

'Well, you may 'ave broken your muvvers' 'earts, but you won't break mine! Now, when I says "On Parade" I wants to see a cloud of dust and a living statue! You gets fell in over there, in three ranks, facing me! ON PARADE!!'

We rushed in dismay to our appointed place in the quadrangle and our worthy corporal began to knock some of the priggishness out of us. For four months we drilled and drilled, every morning, the afternoon being devoted to lectures. We did not resent or even dislike our corporal, who was never unfair or basically malicious, and in fact we collected his imprecations like gems to be treasured.

'Am I 'urting you, lad?' he would inquire of some recruit when inspecting us closely in the morning.

'No, corporal,' came the quavering reply.

'I thought I was treading on your 'air,' he would say. 'GET YOUR 'AIR CUT!'

Or when we were marching and countermarching, a roar would come from him:

'That airman, centre file, fourth from rear. If you don't swing your arm proper, I'll tear it off and beat you over the 'ead with the soggy end!'

Another of his habits was to pick on some unfortunate recruit who failed to put the right amount of energy and precision into his drill and say to him wonderingly:

'You ain't human. You ain't exactly an animal. I'll tell you what you are, lad. YOU'RE THE MISSING BLEEDING LINK!'

Soon after our arrival, we were all subjected to a further series of injections. The medical officer adopted the curious practice of inserting the needle of the syringe in our chests instead of the arm, and the process was rather painful. The following morning we all woke up with red and sore bosoms, some of us feeling dizzy, and were required to parade on the drill ground in thick snow, wearing greatcoats and full packs. One by one the recruits fainted, grey-blue bundles in the snow, whilst the remainder continued to march and countermarch. More and more recruits collapsed until I could hardly refrain from bursting out laughing. The scene reminded me of

A group of young pilot officers being instructed in the use of their new gas masks by a corporal.

Author in Tiger Moth at Nkomo, Southern Rhodesia, April 1943.

something familiar; with a few bloodstained bandages and smashed gun limbers it would have been very similar to a picture that hung in my grandmother's drawing-room entitled 'Napoleon's Retreat from Moscow'. Eventually, even our corporal had to give up.

'Oh my Gawd,' he said. 'What a bunch of bleeding milksops. Unless you're all getting too much of what I'm not getting enough of! FLIGHT, DISMISS!'

On some days we would be sent on long route marches, wearing full packs, threading through country lanes with a squadron leader marching ahead, perhaps eight or nine flights, each of thirty recruits. I enjoyed these marches in the crisp air of East Anglia and vastly preferred them to drilling. The climax came at the end, when the squadron leader would require us to don our gas masks and double-march the last half-mile or so into camp so that we arrived gasping with red and perspiring faces, the eyeholes of our gas masks completely steamed up. The purpose of this exercise was never explained to us, but perhaps the authorities wished to find out if we could run away faster than an advancing gas cloud.

In the lecture room we were introduced to the mysteries of air navigation, morse, meteorology and armoury. The quality of the instruction was not high, for doubtless there were insufficient lecturers to cope with the flood of trainees. Best remembered was a corporal who taught armoury to us, demonstrating with a Browning ·303 machine-gun and requiring us to learn the action of the mechanism by heart. The finale of his description of the action of the breech block mechanism when the gun was fired brought the deathless phrase:

'The bullet nips smartly down the barrel, 'otly pursued by the gases.'

I found the lectures surprisingly simple, although my early scholastic record had been far from brilliant, but learning is usually easy when one wants to learn. I passed the examinations with ease, although some trainees experienced difficulties. We were all desperately anxious to have done with the course and to move on to flying training. It irked us to hear of the German attacks and the defeat of our forces whilst we were wasting time in such futile circumstances. But the flying schools were under great strain and

Probably the best known of all elementary flying trainers, the Tiger Moth was regarded with affection by all war-time 'U/T pilots', including those who failed to qualify such as the author.

could not cope with the flood of trainees, and the air training schools in Canada and Rhodesia had barely begun to operate. For me, one of the more pleasant aspects of this period was to make so many new friends from all parts of the country, and from such diverse places as Australia, Canada and South America. We were all so young and keen to fly, and most of us considered ourselves invulnerable. Had we known what lay ahead our enthusiasm might have been blunted, for only a handful were to survive the war.

On 19th May 1940 we were posted at last, and I found myself in Upwood aerodrome, near Ramsey in Cambridgeshire; but to my intense disappointment this was not a training school and we spent most of our time sitting in machine-gun pits, our circumstances being enlivened by a real air raid during which sixteen bombs were dropped on us, causing one death; my gun pit was narrowly missed, although I was not in it at the time. However, our posting to flying training arrived in the middle of June, to Prestwick in Ayrshire.

My flying training began on 18th June. The days of Dunkirk and the capitulation of France were now behind us, and all of us realised that the war was likely to be long and grim, if the UK was to survive at all, and that we would be in the front line of our defences. Our training aircraft was the Tiger Moth, that delightful bi-plane which looked like a survivor from the Great War but which was safe and a pleasure to fly, provided one had the aptitude. My trouble was that I did not seem to possess the right 'air sense', and that immense flying skill that should have been my birthright failed to materialize. After a couple of weeks and about eight hours' instruction I went solo and continued both dual and solo for another eight hours' flying.

In this time many hopeful trainees had been suspended from the course and on 11th July I was advised to switch to an air observer's course, and decided to accept this even though I was given an opportunity to continue pilot training. In spite of the severe blow to my ego, I realised that my natural aptitude as a pilot was limited and might never develop sufficiently; on the other hand I found the classroom subjects such as air navigation extremely easy to understand, and I was consistently top in all subjects. It was these results which prompted the Chief Instructor to recommend my transfer, for the RAF had at the time an even greater need for observer/navigators, to release the second pilot who at that time acted as navigator in bomber aircraft, and so to supply at once an experienced man to act as first pilot. After a brief spell in Babbacombe, near Torquay in Devon, where a group of us fulfilled the unlikely roles of service policemen, I was back in Prestwick on 11th August 1940 to begin my air navigation course. This lasted twelve weeks and the work was far more intense than anything I had experienced, but I took to it like a duck to water.

My first flight was called Air Experience. Thirty trainees entered a four engined Fokker passenger plane, acquired from Dutch Air Lines, and flew on a bus ride around Western Scotland, learning how to identify ground features from maps. It was an enjoyable experience, and most amusing to begin our training in a Dutch aircraft. After about twenty flights on these Fokkers, the remainder of our flying was in the Avro Anson, a twin-engined monoplane which was to see long service in the RAF in training navigators. There were two trainees as well as the experienced pilot, and each trainee took turns in calculating the course and time of arrival to a particular objective, taking into account such factors as the airspeed of the aircraft, the estimated wind speed and direction, the magnetic variation of the earth and the magnetic deviation caused by the aircraft itself. This procedure was called 'Dead Reckoning'. The other trainee checked the position by map reading when the ground was not obscured by cloud, took compass bearings of landmarks and calculated the drift of the aircraft by linking up the ground with the bombsight. Aerial photography was also practised. Even flying at the moderate airspeed of 120 knots, these calculations were a constant race against time.

To my humiliation, I felt airsick at first, particularly when lying on my belly looking down the bombsight in bumpy weather, but this uncomfortable feeling passed away after a week or two. Throughout the war, I always felt airsick after resuming flying after a longish period on the ground, but always recovered. At the end of the war, flying in Dakotas over Burma in violent monsoons had no effect on my stomach, although all passengers and some aircrew were unpleasantly sick. On the ground, instruction continued apace, covering such subjects as navigation, compasses, meteorology, maps and charts, aircraft instruments, reconnaissance and photography. The quality of the lecturers was excellent but it was very hard work for the trainee, and all waking hours were occupied if one wished to pass the course. Examinations took place regularly and those who failed were ruthlessly weeded out.

Halfway through the course, towards the end of September, we were given a long weekend leave. It was less easy for me to reach home at Woodford Green in Essex than when I was stationed in Cambridge, but I boarded a train for the long journey from Glasgow to St Pancras. The train was much delayed and as we neared London an enormous rosy glow lit the night sky, punctuated by searchlights and the vivid explosion of bombs. We were witnessing a major attack, but by the time the train drew into the station the all-clear had sounded. London seemed ablaze with fire and I was fortunate in finding a taxi and persuading the driver to take me to Liverpool Street.

'I'll do my best, mate,' he said, but as we approached St Paul's we were met by the crumbling walls of burning buildings and were motioned away by angry wardens and firemen.

'Can't drive over the hose pipes,' said my imperturbable driver. 'We'd better turn back.'

But I paid him off with profuse thanks, shouldered my kitbag and began to pick my way through the burning area of devastation. I was skirting past streets that had been familiar to me barely a year ago but now were choked with mounds of smoking rubble. In Newgate Street, Christchurch lay in ruins, where generations of my mother's family had been christened, married and buried. Thousands of feet above, the young men of the Luftwaffe, diverted by Hitler's direct order from attacks on RAF aerodromes to open the blitz on London, had pinpointed their positions over the meandering Thames and released their loads over the docks and the City, mangling bodies and destroying in seconds the buildings and artefacts of centuries. Their Heinkels, Dorniers, and Junkers had now turned for home, leaving some of their number behind. The luckier fliers would have been killed quickly, but some would have crawled with smouldering parachutes along twisting and flaming fuselages, shocked or injured, in desperate attempts to wrench open escape hatches before the

Fokker F.XXXVI G-AFZR at Prestwick aerodrome. Only one of these aircraft was built, and it was sold to a British company in 1939 and then acquired by the RAF as a navigational trainer.

The author walked past scenes similar to this when returning home on leave in September 1940, when the City of London had suffered a very heavy raid from bombs and incendiaries. A constant anxiety for aircrew with relatives in London and other major cities was the fear of bombing raids in the early part of the war.

Fairey Battle.

The Fairey Battle was obsolescent when the author was under training in 1940, being employed for training air gunners and bomb aimers. Two trainees would take turns at firing the Vickers K at a drogue towed behind another Battle, or lying prone behind a course-setting bombsight in the belly of the aircraft and dropping practice bombs on rafts moored in the sea. In the photograph the trainee is aiming a Lewis machine gun and not a Vickers K.

impact of earth or water ended their agonies; a spectre of death that was to grimace at me for the next eighteen months.

Around the fiery ruins, purposeful firemen, air raid wardens, nurses and policemen, were working like demons on the edge of a holocaust. The sturdy competence and bravery of my fellow Londoners amazed me. 'If these people can take punishment in this way, then perhaps I can cope with the dangers ahead,' I thought to myself. Nevertheless I was in a ferment of fear for the safety of my parents, for not only did my father work by day in the City but the blitz might have extended to the suburb where they lived. But as the train from Liverpool Street approached our local station it was a relief to find that our district was almost unscathed, and characteristically the family seemed cheerfully impervious to the threat from the air. My parents and two younger brothers lived in a detached suburban house and had constructed a splendid underground shelter in the garden, where they often spent the nights. There always seemed to be a raid when I went home on leave, however, and the danger of a direct hit worried me seriously. This possibility receded somewhat in 1942/3, but recurred in 1944/5 with the advent of the V weapons.

My air navigator's course at Prestwick ended on 15th November 1940, and I was gratified to find that I had achieved the best results. To commemorate the occasion, a group of us went into Ayr for a party of celebration. We mixed our drinks in an amateurish way, and for the first and only time in my life I became sick drunk. I can vaguely remember vomiting in a churchyard and being helped on to a bus by a group of solicitous airmen, who thoughtfully paid my fare out of my wallet. They also removed the remaining contents for their own benefit, as I discovered on waking with a pounding headache next morning, and this was another lesson that I did not forget.

Another incident which puzzled and amused me was that whilst on this course I was sometimes saluted by airmen who had recently arrived from Yugoslavia. As a leading aircraftman, I was entitled to salutes from no one, albeit most of us had had our new uniforms re-tailored privately and wore a white flash in our forage caps to denote our special status as cadets. 'I'm obviously officer material,' I joked to my companions whilst graciously returning these salutes, but thought that the Yugoslavs could not recognize our lowly rank.

The explanation came two years later in Southern Rhodesia when I walked into the officers' mess at Cranborne and a group of senior Yugoslav officers leaped to their feet. Passing them by, I was ordering a drink at the bar when the most senior, a group captain, approached me and bowed, addressing me deferentially in an incomprehensible language. It appeared that I bore marked resemblance to King Peter of Yugoslavia, who had been forced to leave the country when it was overwhelmed by the Germans. I wondered how I could turn this surprising revelation to my advantage after the war, but my doppelganger was eventually deposed and became a US citizen.

From Air Navigation school the next move was to Bombing and Gunnery school, which was held at Stormy Down, Porthcawl, in South Wales. In the early days of the war, the navigator had to perform the function of bomb aimer as well as man a gun turret in an emergency. Once again, the course consisted of an intense combination of ground subjects and air exercises. I already had a good knowledge of the Browning and Vickers machine-guns but had never fired either in practice. At first we fired at targets on a range and I found this a pleasurable experience, for my left-handedness was of no importance and it was highly satisfying to tuck the butt of a machine gun under the chin and let forth bursts of fire at over 1,000 bullets per minute. This was followed by more ground practice, in a gun turret mounted inside a black dome, where we learned to follow a light moving around the interior and to practise deflection shots at model aircraft from different angles, using the ring and bead sight and simulating firing.

In the air, our gunnery practice took place in an obsolescent aircraft called a Fairey Battle and also the Armstrong Whitworth Whitley. Both of these were flown mainly by Polish pilots. The trainees stood behind the drum-fed Vickers in the rear cockpit of the Battle, or sat behind the belt-fed Brownings in the mid-under or rear turrets of the Whitley and fired bursts of tracer at a drogue towed by another aircraft attacking from various angles. Afterwards the number of holes in the drogue were counted and the performance assessed.

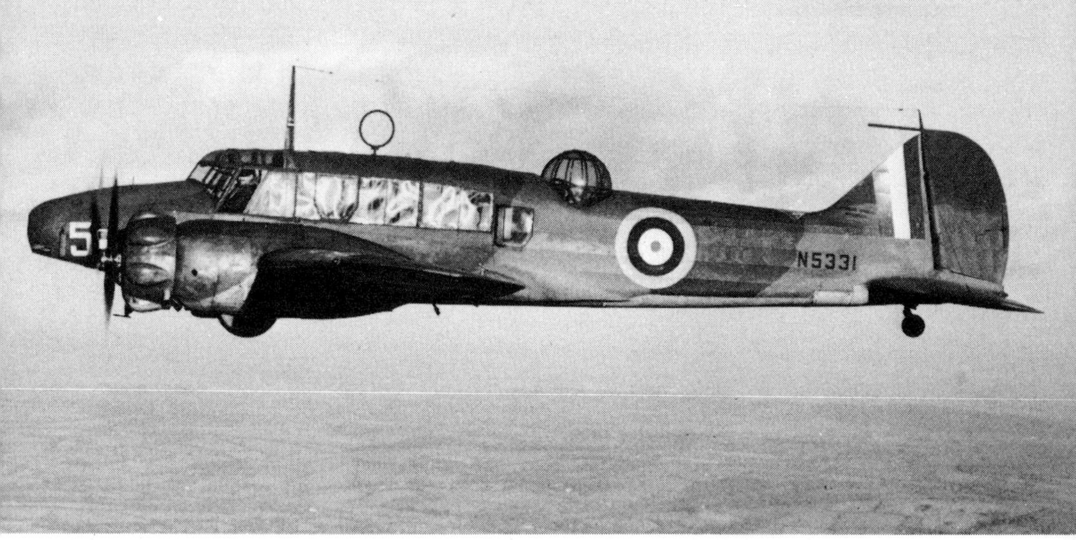

In 1940 the Avro Anson was widely used for training navigators. There were usually two trainees in the aircraft, one practising with the 'course setting bombsight' and map reading, the other working on the chart table with 'dead reckoning'.

The Armstrong Whitworth Whitley before the rear turret was fitted.

The rear turret of the Armstrong Whitworth Whitley, armed with four Browning ·303 machine guns.

Bombing practice took place over bright yellow rafts anchored in the sea. To achieve accurate results, it was necessary to calculate exactly the wind speed and direction which affected the bomb during its period of falling and which had to be set on the bombsight. As a very rough measure, the time a fully streamlined bomb takes to fall in seconds is a quarter of the square root of the height in feet so that, say, from a height of 10,000 feet the time would be about 25 seconds. Thus an error of 5 mph in wind speed alone could cause an error of nearly 200 feet from this height, no matter how accurate the bombsight.

The bombsight was called a course setting bombsight. It needed to be levelled with spirit levels and on it the bomb aimer set the wind speed and direction as well as the aircraft's height and speed, and the terminal velocity of the bomb. From the bombsight protruded a 'drift bar' of about a foot long, and this bar altered direction slightly with these settings. As the aircraft approached the target the bomb aimer gave instructions to the pilot such as 'left, left, right, or steady', until the target passed down the drift wires and appeared in a ring sight, when he pressed a button and released a bomb, saying 'bomb gone'. Later in the war, an improved bombsight called the Sperry Gyro was introduced; this levelled itself automatically and was a great improvement, for the CS bombsight was inaccurate if the aircraft was not flying dead straight and level.

As the practice bombs fell near the target the splashes and small explosions were plotted from the shore and, again, performance could be assessed.

My results in these air tests were satisfactorily high, as were my pass marks in the written examinations and in morse. The course lasted for about eight weeks, and we lived in extreme discomfort, being billeted in a great barn-like amusement arcade on Porthcawl pier. As usual this was unheated and we slept wrapped in newspapers and blankets whilst the wind whistled through the floorboards and wooden walls during another bitter winter.

At the end of the course, six of us were commissioned as Pilot Officers, General Duties, whilst the remainder who passed became sergeants. We had no choice as to our next posting and the next move should have been to an Operational Training Unit, but three of us who had achieved the best results and also been commissioned, 'Jock' Maclean, 'Monk' Matthews and myself, were posted directly to an operational squadron, number 217, based at St Eval near Newquay in Cornwall. We read the news of our posting on the notice board with pleasure not unmixed with apprehension for by now we had a fair idea of the dangers that awaited us. We made some enquiries about 217 Squadron. It proved to be a reconnaissance squadron of Coastal Command, and its squadron crest bore a shark's head protruding from the sea, with the motto 'Woe to the Unwary'.

CHAPTER TWO

Through the Balloon Barrage

'To be conscious that you are ignorant is a great step to knowledge.'

BENJAMIN DISRAELI (1845)

Although I had become an officer, I was without a uniform, having returned my aircraftman's kit to the stores at Porthcawl. The battledress which became common amongst aircrew had not yet been introduced. A London tailor with an eye to business had measured the commissioned ex-trainees at Porthcawl immediately their appointments were announced and I had purchased such items as hats, gloves, greatcoat, shirts etc., but the jackets and trousers were made to measure and could not be ready for a few weeks. I therefore reported to St Eval late one cold and wet January evening in 1941, a nineteen-year-old feeling highly self-conscious in civilian clothes.

As so often seemed to happen, my arrival was heralded by a bombing attack. St Eval seemed to have a permanent arrangement with Morlaix in Brittany, for squadrons in the two aerodromes raided each other frequently, and that night it was the turn of the Germans. The Luftwaffe had a slight advantage at night, for their Heinkels could map-read their way up the north coast of Cornwall until they reached the easily identifiable landmark of Trevose Head and then turn on a set course south-east for St Eval. They could then release their bombs by timing two minutes or so and be pretty sure of hitting something on the large but darkened airfield.

On this occasion the guard at the main gate told me that they had scored a direct hit on an air raid shelter, killing a number of airmen and airwomen. The station was alive with ambulances and fire engines as I made my way to the officers' mess on that late evening, to find myself a meal and a bed for the night. The following morning I went to the stores and drew out my flying kit. This included a

A view of St Eval aerodrome taken in March 1941 and looking towards the Cornish coast. Note the camouflage effect on the runway. The aerodrome was being subjected to constant air attacks from German-occupied France.

A view looking from St Eval towards Newquay. The indented coast provided excellent recognition for enemy and friendly aircraft alike.

Sidcot suit, a one-piece garment covering the body and legs, with a separate inner lining of quilted cotton which could be attached with press studs to the outer portion in cold weather. The Sidcot was fitted with numerous zips and fastenings and was rather a cumbersome piece of clothing. Far more popular was the Irvin jacket and trousers, sheepskin articles which were more comfortable, warmer, easier to put on and looking more professional.

I succeeded in drawing all these articles, together with helmet, flying boots and a Mae West lifejacket. Then I went to the armoury for a ·38 Smith and Wesson six-round revolver and lastly to the parachute section for a harness and parachute pack. Thus equipped with these necessary articles hiding my civilian clothes I boarded a truck to 217 Squadron crew room near a small church on the opposite side of the airfield. No one seemed in the least impressed with this new arrival and the only spark of interest was from a sergeant pilot who asked if I knew how to handle a power operated turret, since he had to take off shortly and his air gunner had reported sick.

I responded enthusiastically and joined the crew in walking out to their Bristol Beaufort, an aircraft which was unfamiliar to me but which I had been eyeing curiously for an hour or so. The mid-upper turret, which I was to occupy, traversed sideways and rearwards and was armed with a ·303 Vickers K machine-gun. Happily ensconced in this warlike position, I was thrilled when we took off and began a daylight patrol looking for U-boats near the south coast of Ireland.

To my dismay, however, my turret rapidly developed a fault and would only traverse to aft and starboard, which would have been somewhat awkward if the Luftwaffe attacked us on the port beam or quarter. However, I was too shame-faced to report this defect to the pilot, especially since it might have put an ignominious end to my first operational flight. The work of scanning the sky for aircraft and the sea for U-boats for about four hours proved surprisingly onerous. It was difficult to maintain continuous concentration and I had to discipline myself to sweep the foreground and background of the sea in a set pattern and to keep refocusing my eyes in the sky for signs of aircraft.

Disappointingly for me, nothing was seen, but I did not yet realise that this was normal on such patrols. The hazard came from aircraft failure or air crew error rather than from the enemy. On the homeward leg I reported the turret defect to the pilot and to the armourer on return. This flight was not recorded in my log book, however, in case I got into trouble for flying in civilian clothes.

Number 217 'Exeter' Squadron was formed in 1914, disbanded in 1919, and reformed in 1937 for general reconnaissance work in Avro Ansons. At the time I joined the squadron, the crews had recently finished converting to the new Bristol Beaufort. Most of the pilots were 'short-service' commissioned officers, or non-commissioned officers of the rank of sergeant and above in the regular RAF. They were all experienced fliers, many of them having seen action in other aircraft such as the Anson or the Bristol Blenheim. There were only a few navigators already on the squadron and the role of the newcomers was to release the second pilots from their function as navigators so that they could become first pilots.

At this time, there were two more experienced squadrons which had been fully operational on the Beaufort for some months. These were numbers 22 and 42 and the crews had converted in 1940 from the Vickers Vildebeest, an ancient torpedo bomber, to the new Beaufort which was intended to take over that hazardous role. The crews on 217 were not torpedo-trained, however, and it was intended that we should use the new Beaufort as a dive bomber or for low level attacks on enemy shipping, naval vessels or ports. In addition, the squadron could still be used for the rather mundane work of convoy escort, reconnaissance patrols and U-boat hunting as well as the more dangerous work of mine-laying directly outside the entrance to the more accessible European ports. Our sorties could thus vary from the almost suicidally dangerous in the form of a low-level daylight attack on a warship to the fairly safe but exhausting work of circling around a homeward bound convoy in the Atlantic Ocean.

The Bristol Beaufort 1 was originally designed for the RAF as the world's most modern torpedo bomber and production began at Filton, near Bristol, just before the war. It was of the same family as the Blenheim but slightly larger with a wing-span of 58' and length of 44'. The Taurus engines of 1,130 hp powered this heavy all-metal mid-wing monoplane. The cruising speed was about 165 mph (about

ernal views of the Beaufort
the pilot's seat — note
ton on left of control
umn for firing the Browning
3 mounted in the port
g.

pilot's seat, looking
ards the sergeant
igator, seated by his chart
e on the nose. The
ufort was so noisy that
t and navigator had to use
intercomm, even at this
rt distance

143 knots) and the maximum speed around 270 mph (about 234 knots) the endurance being about six hours. It could carry 2,000 lb of bombs or a torpedo, landmine or seamine of roughly equivalent weight.

The Beaufort carried a crew of four. The pilot sat in the left-hand front seat, looking over the nose. The navigator sat beside him on take-off or landing, but his 'office' was in the perspex nose, from which he had an excellent view. Our aircraft had one forward-firing Browning ·303 mounted in the port wing and operated by the pilot, whilst the navigator was equipped with a similar machine-gun mounted in a 'blister', a perspex turret under his seat; this was a backward-firing gun intended to protect the aircraft from a belly attack and was fired from a prone position looking through a mirror, but the cartridge belt constantly fouled in its chute and later this gun was removed.

The navigator sat on a revolving seat so that he could work on his chart table on the left, whilst from a slightly different position he could look through his bombsight in the nose. He also had various instruments such as an automatic bomb distributor, nicknamed 'Mickey Mouse', an Aldis lamp for signalling, and a bearing compass. Behind the pilot was an armourplated bulwark, then the radio equipment and the wireless operator's seat. The wireless operator was also an air gunner and carried a Vickers K machine-gun which he could mount in the hatch during an attack. Lastly, there was the air gunner, who sat in his power-operated turret with its two drum-fed Vickers K guns. Originally the Beaufort's turret was equipped with only one drum-fed Vickers K gun, but some of our aircraft were fitted with two, and later in 1941 were equipped with two belt-fed Brownings.

The crews in our squadron were not happy with their new aircraft, for the Beaufort was difficult to fly and had already acquired a bad reputation for mechanical failure. The main criticism concerned its inability to fly on one engine. Moreover, the crews thought that it sank like a stone if the pilot was forced to land in the sea, so that there was no time to release the dinghy housed in the port wing. We all knew that we could rapidly die of exposure bobbing about the Atlantic in our Mae West lifejackets. Several Beauforts had

The blister gun, fired by the navigator, looking through a mirror, was intended to protect the Beaufort from a belly attack, but the cartridge belt of the Browning ·303 constantly fouled in its chute. This photograph was taken in early 1941 at Leuchars, and shows a 42 Squadron Beaufort. The central figure is F/Lt Simmonds who later became S/Ldr in 217 Squadron and completed over 100 operational flights in all, and was awarded the DFC.

The early Beauforts were equipped with a single Vickers K·303 machine gun in the mid-upper, or dorsal turret. In 1941, two Vickers Ks were installed in the turret.

In 1942, the Vickers Ks were replaced with two Browning ·303s in the turret. A Vickers K was installed in the waist hatch in 1941.

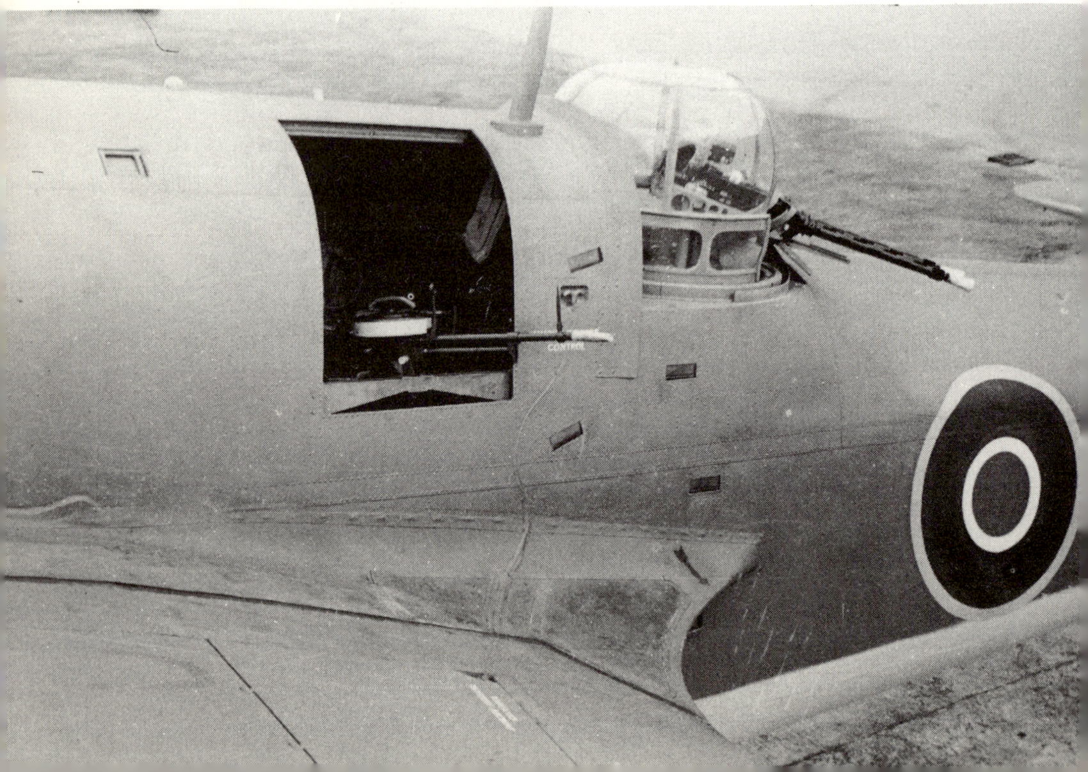

unaccountably disappeared from flights over the sea, and for a time the aircraft had been grounded whilst the Taurus engine was modified. Another feeling amongst torpedo pilots was that although the Beaufort was considerably faster than its predecessors, there was no improvement in the torpedo which still had to be released at a slow speed, so that aircraft was an easy target for a ship's gunners during the lengthy run-up when attacking.

As yet, I had been unable to join a crew and the month of February was occupied with only the occasional flight for familiarization. St Eval was a large station, and there were squadrons of Lockheed Hudsons, Wellingtons and Bristol Blenheims operating from it, as well as our Beauforts. My uniform arrived and I settled down in the officers' mess on the station as well as in our reasonably comfortable billets in the Watergate Hotel, built on top of a steep cliff overlooking a bay north of Newquay.

By the second week in March, however, I crewed up with a young pilot, a flying officer with a short-service commission in the regular RAF and we flew on a couple of convoy protection flights south of Ireland, both of which were of little value since the weather was atrocious and visibility almost nil. But I was very happy with my new pilot, who was extremely likable and seemed both competent and steady, whilst my dead-reckoning navigation proved accurate enough.

My first taste of action came in the middle of March when our crew was on stand-by in the operations room and a report came in concerning some British merchant vessels that were being attacked by German aircraft whilst south of Ireland and heading for the Bristol Channel. We flew off to investigate and came across a burning tanker in the St George's Channel between Wales and Ireland. We were flying in difficult conditions with low cloud and poor visibility, and the stratus cloud kept covering the unfortunate tanker so that we could only see it intermittently. To check the position and in order to signal base we headed south to the Smalls Lighthouse, west of Pembroke, where we spotted two rubber dinghies in the sea. Circling low, we could see that they were tied together and that five men dressed in flying suits were clinging to the lifelines inside. We quickly sent a message back to base.

The sea was very turbulent and the dinghies were riding badly in the waves. I flashed with the Aldis lamp to the airmen and we all waved to them, but they did not respond at all although they were looking directly at us. The dinghies were a strange shape, being large, orange and squarish, whereas the RAF dinghies were yellow and circular. I was looking at them carefully through my RAF binoculars. Suddenly we realised that the survivors must have been the five-man crew of a German Heinkel, and perhaps they were wondering if we were about to machine-gun them. Of course we would never have attacked downed fliers in that condition and indeed we felt considerable respect and almost no antagonism. Often, we used to say that we should like to talk to our adversaries and when I did this after the war I found that they were not dissimilar to ourselves.

The weather was clearing a little, and through a gap we spotted a small merchant vessel heading towards the Bristol Channel. We flew towards it, again circling low. I signalled in Morse, using the hand-held Aldis lamp 'Please pick up men in life raft'. They understood, but some seamen lining the rail gave us a thumbs down whilst somebody flashed back from the bridge 'No bloody fear'. Perhaps the rocks are too dangerous for a rescue attempt, we thought, or perhaps this was another ship that the Germans had attacked, and we flew back to the dinghy and resumed circling.

By now the dinghies were being swept on to the rocks of Smalls and suddenly they reared up and a wave swept over them. A few seconds later they reappeared with only three men clinging inside. Again and again the waves swept over and the remaining three were dragged one by one into the water, so that at last there were the empty dinghies swirling in the frothy white water and we could see no sign of the airmen.

Sickened, we turned for home. It might have been us down there, we thought, and a feeling of gloom and helplessness overcame us, so acute that we could barely discuss the matter amongst ourselves.

Another Beaufort, piloted by Pilot Officer Dunn, arrived on the scene just after we left. The two empty dinghies were still there, and a body could also be seen in the water. Dunn tried to attract two naval vessels to the rescue, but they refused, since they had recently been attacked by Heinkels. At this, three Heinkel 111's appeared and Dunn tried to fire at them, but two of the German bombers vanished in the low cloud and the third weaved low over the sea to escape. Yet another Beaufort, piloted by Sergeant Routledge, spotted a Heinkel 111 crashed and burning on Lundy Island, and this may have been connected in some way with this series of attacks.

The weather in that dull spring of 1941 was continually wet, cold and unpleasant, so that much of our flying was cancelled. We were frequently briefed for sorties that did not take place and we could often stand on the Cornish cliffs watching the sea fogs or rain squalls approaching us from the Atlantic. On one gloomy overcast day we were sitting in our crewroom near the perimeter track, smoking and playing cards, when we were asked to help push one of our Beauforts which had become stuck in the mud. We all trooped out and drifted in a body towards the immobilized aircraft when we heard the noise of unsynchronized engines and a twin-engined bomber emerged from the low clouds some distance away. We all looked up at an unfamiliar shape and someone said, 'Is that one of those new Beaufighters?'

At this the bomber dipped its port wing and banked towards us, revealing its outline.

'Heinkel!' we yelled, and as one we turned and bolted.

At school, I was more of a cross-country runner than a sprinter, but my performance on that occasion would have constituted a school record, even encumbered by flying kit. I seemed to bound through the air with legs pumping, like a character in a Walt Disney cartoon, whilst from behind us came the increasing roar of the aircraft and the rattle of its machine-guns. In that sprint, I seemed to be leading the field but, as we crossed the perimeter track and leaped into the drainage ditch beyond, my boots landed squarely on the back of an airman who had already taken refuge there. He turned round to curse, but we were enveloped by flying bodies as the remainder of our party joined us in that water-sodden ditch.

For some reason, we were helpless with laughter and rolled about with great hilarity shouting, 'That was a bloody fine Beaufighter!' Emerging a few minutes later, we found that no one had been hit, but our clothes needed drying.

For the newly formed crews on 217 Squadron, it became evident that we would soon be engaged on more serious flying. Convoy work continued, each of my flights being uneventful, but these were interspersed with practice dive and low-level bombing. Our range was usually Trevose Head, a string of rocks jutting out into the Atlantic, and we perfected our techniques with small practice bombs that exploded with a puff of smoke. The pilot would dive at a fairly shallow angle and release the bombs singly, each of which seemed to hit the target. Alternatively we would scream in at a low level and the navigator would release the bomb, often by visual judgement since the bombsight was of less use at such a height. We became extremely proficient with this practice. Sometimes we would use a sea marker as a target instead of the rocks, this being a bag of aluminium dust which we dropped down the flare chute and which spread so that it formed a silver splodge on the sea, easily seen from afar. On one such occasion, a clumsy wireless operator dropped one of these bags inside an aircraft so that the silver dust spread in a thick cloud throughout the fuselage, enveloping instruments and crew alike. A partial clearance was made by opening a hatch and the aircraft landed to disgorge the four crew looking exactly like silvery ghosts.

The navigator was Pilot Officer 'Monk' Matthews, a large and stolid countryman from Trowbridge, Wiltshire, who was a little older than most and lived with his wife in a small cottage near the Watergate Hotel. His clothing and skin were completely coated with the dust and although we were inclined to laugh at him at first, we drove him for a bath. Matthews got out of the car and began to walk down the rather long path to his front door, but his young wife was looking out of the window and nearly collapsed when she saw this ghostly apparition. Probably she was in a permanent torment of anxiety for her husband's safety, and it often struck me as an unsuitable arrangement that wives should live with men who risked their necks daily in our dangerous pursuits. The strain on the nerves must have been more acute than amongst the single men who had only themselves to worry about.

The crew of this aircraft were quite unwell after this episode for the dust had penetrated their lungs and stomachs and gave them a very unpleasant but temporary attack of poisoning.

A Beaufort crew of 42 Squadron putting on flying kit and Mae West life jackets prior to an attack on the Prinz Eugen off Norway in May 1942.

A Heinkel 111 is shot down into the Atlantic with little chance of survival for the crew.

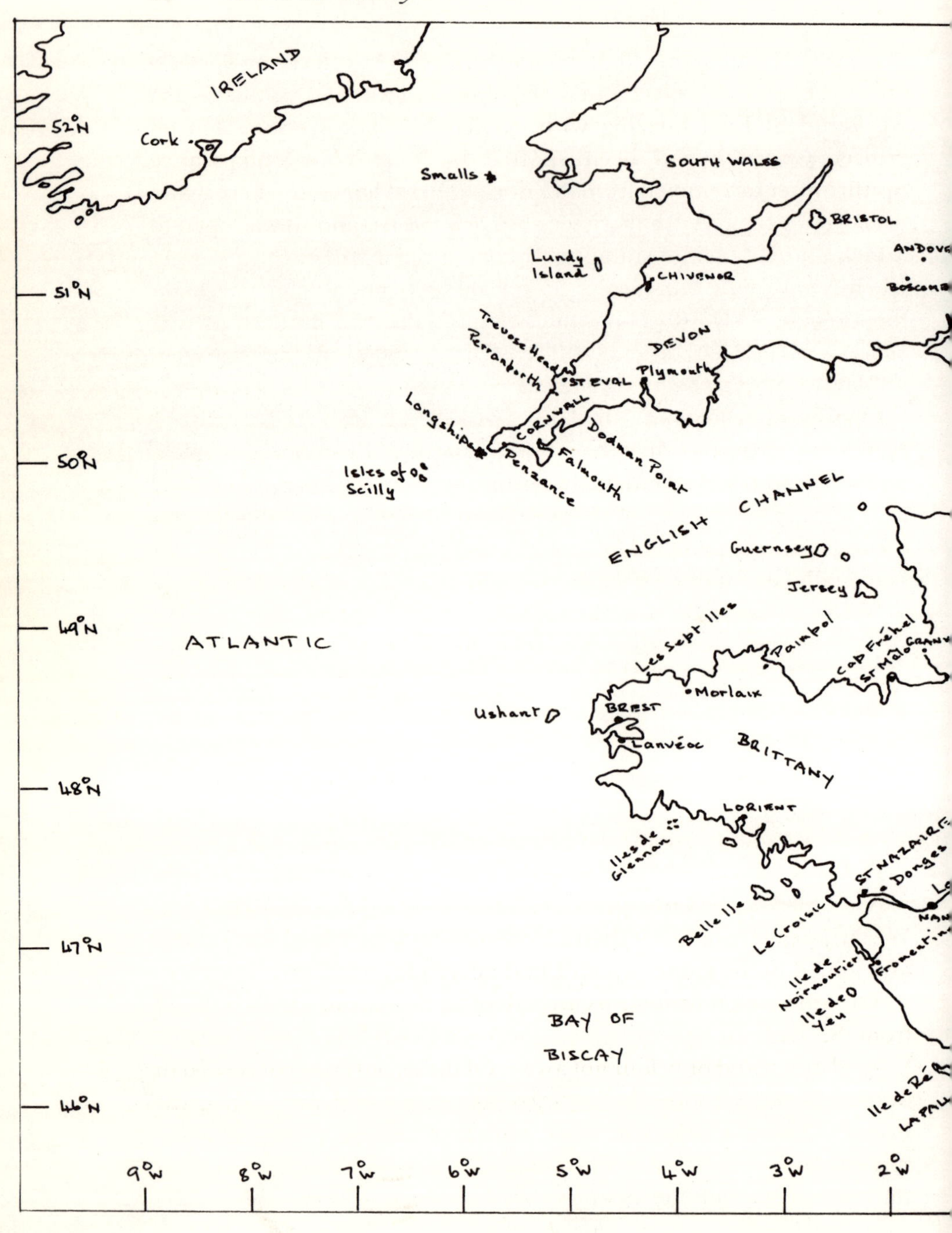
52°N
51°N
50°N
49°N
48°N
47°N
46°N
9°W
8°W
7°W
6°W
5°W
4°W
3°W
2°W
IRELAND
Cork
Smalls
SOUTH WALES
BRISTOL
ANDOVE
Lundy Island
CHIVENOR
BOSCOM
DEVON
Trevose Head
Perranporth
ST EVAL
Plymouth
CORNWALL
Longships
Dodman Point
Penzance
Falmouth
Isles of Scilly
ENGLISH CHANNEL
Guernsey
Jersey
ATLANTIC
Les Sept Iles
Paimpol
Cap Fréhel
St Malo
GRAN
Morlaix
Ushant
BREST
Lanvéoc
BRITTANY
LORIENT
Iles de Glénan
ST NAZAIRE
Donges
Belle Ile
Le Croisic
Ile de Noirmoutier
Fromentine
NA
Ile de Yeu
BAY OF
BISCAY
Ile de Ré
LA PAL

In early April I had to navigate a newly commissioned but highly experienced pilot named Kitching to Squire's Gate, ferrying another crew who were to collect a new Beaufort aircraft, and on our return we found St Eval buzzing with excitement. The German battle-cruiser *Gneisenau* had been spotted by a photo-reconnaissance Spitfire after its removal from dry dock in Brest harbour where it had been under repair with its sister-ship *Scharnhorst*, and so was open to attack and a possible sinking. Of course, we had been wondering when it would be our turn to attack these formidable targets as in their refuge at Brest. The name we gave the two battleships was *Salmon and Gluckstein*, after a well-known catering firm, but they were unlikely to prove a tasty dish for us.

On the afternoon of 5th April I wandered into 217 Squadron crewroom, to find it full of the more experienced Beaufort crews of 22 Squadron who were on detachment from their station at Chivenor, near Barnstaple. Some of them seemed to be writing intently and sorting out their personal belongings, but I did not know any of them personally and asked a friend:

'What on earth are they doing?'

'Writing out their wills of course,' he said. 'They're going on a suicide mission!'

Deep in thought, I drifted out again, but suddenly heard my friend running after me.

'Hide!' he shouted. 'They're looking for spare navigators!'

'Shouldn't we volunteer?' I asked, hesitatingly.

'No, it's not our squadron,' he replied. 'Follow me!'

In the distance, on the edge of the airfield, was a group of airmen playing hockey, and my quick thinking friend led me towards them. We stripped off our incriminating officers' jackets, seized some sticks and joined in the game, which I had never played before, nor since.

The following morning, six aircraft of 22 Squadron tried to take off from St Eval in appalling weather conditions, but two became bogged down and only four got away. Of these, only one succeeded in finding Brest harbour through the low cloud and rain. This was Flying Officer Campbell, who somehow managed to drop his torpedo between the mole and the *Gneisenau*, which nearly sank from the explosion. But the Beaufort was blown out of the sky by the massed defences, which was reputed to be the heaviest in Europe, and crashed into the harbour; all four crewmen were killed. Campbell was awarded a posthumous Victoria Cross.

Years later, I read a German account of this episode, which was full of admiration for the crew, but asserted that the navigator was at the controls when they dredged the Beaufort out of the harbour. This was a Canadian called Sergeant Scott, who was awarded a posthumous Distinguished Flying Medal, and he had probably managed to take over the controls for a few seconds after Campbell had been killed or wounded. The Germans buried the crew with full military honours.

Thoroughly ashamed at my secret cowardice, I hoped to redeem myself by achieving some feat of bombing. For me, however, the next flight was a search for the crew of a missing aircraft over the sea, from which we had to make an early return when one of our engines began to develop trouble. But at last, I was ordered on a night patrol off Brest harbour, temporarily joining the crew of another short-service pilot, Flying Officer Dunn, which promised to give me the opportunity of seeing some real action. We took off at dusk on 11th April and I was in some trepidation for I had never flown in an aircraft at night before. The Beaufort was dark and cold as I stumbled down the fuselage carrying the tools of my trade in my navigator's canvas bag, before strapping myself into the co-pilot's seat.

Taking off with a full bomb load down the flarepath lighted only by gooseneck flares with a rather unnerving experience for me, but I entered the nose compartment as soon as we were airborne, switched on the Anglepoise light and gave Dunn the course to Land's End. I was relieved to find that I could distinguish the coastline beneath us, but its configuration was already familiar to me. We used a 1:250,000 Ordnance Survey map, coloured and with ground features clearly marked, but it was quite tricky switching a torch off and on the map whilst attempting to pick out landmarks below; nevertheless I seemed to be coping well enough. From Longships, the lighthouse off Land's End, we set course for Ile d'Ouessant, the most westerly point of France and known to centuries of British mariners as the Island of Ushant. To my intense relief, for we had no navigational aids except our magnetic compass and pilot's gyroscope, the island put in its appearance some fifty minutes later, in the right place at the right time.

We were flying at 1,200 feet, and from Ushant our task was to fly up and down Brest Estuary, on what was called a 'Stopper' patrol, to look for and attack any of the German battle fleet making for the Atlantic. However, as we crossed the easterly part of the island, I had my first experience of flak, curving gracefully up to meet us, and shouted to Dunn, 'Tracer coming up! Directly ahead!'

Dunn banked to starboard and the tracer whipped past our port wing. It was quite close, a single stream of flying orange sparks. Then there were some red flashes and I could smell cordite.

'How exciting!' I thought. 'That must be Bofors fire. And I don't feel in the least frightened!' Ignorance was bliss at this stage.

We turned and left this light flak behind us, settling down to our patrol. I was hard at work calculating a series of very small courses, sometimes peering at the dark sea beneath us and the silvery lines of surf on both sides of the estuary. To the east, searchlights and ominous flashes kept us aware of the presence of Brest and the German battle fleet, but our job was to maintain our patrol and we did not approach this formidable target closer than a couple of miles.

Time passed astonishingly quickly, just as it does when one is racing to complete answers in an examination. The weather was clear and the night was starry as we set course from Ushant once more, heading for Penzance. I was feeling very pleased with myself. More than halfway back to the Cornish coast, however, the wireless operator came up the fuselage with a piece of paper. Dunn read it and passed it to me.

'Divert to Boscombe Down' was the instruction from St Eval.

'Boscombe Down? Where the devil's that?' I thought, and hunted in my map.

Eventually I found it, a long way inland, near Salisbury, and I entered the position on my chart, worked out a new course and passed it to Dunn. After we turned, the reason for the diversion became obvious, for low stratus cloud covered the sea and fragmented alto-stratus appeared above us.

'Must be a warm front moving across,' I thought, with my recently (*opposite*) The pilot of an aircraft who found that he had entered a balloon barrage had almost no alternative but to climb and try to fly through to the other side. If he tried to turn back he would be likely to hit cables in the turn.

Entering a balloon barrage was far more dangerous than flying through a town's flak defences. As the diagram shows, if the Beaufort had struck one of the many cables the latter would have parted at top and bottom and two parachutes would have opened at each end. The enormous drag would have sent the aircraft crashing to the ground, whilst the balloon would have deflated slowly.

As an inexperienced navigator, the author foolishly directed this aircraft, L9878, through Plymouth balloon barrage on the night of 11th/12th April 1941.

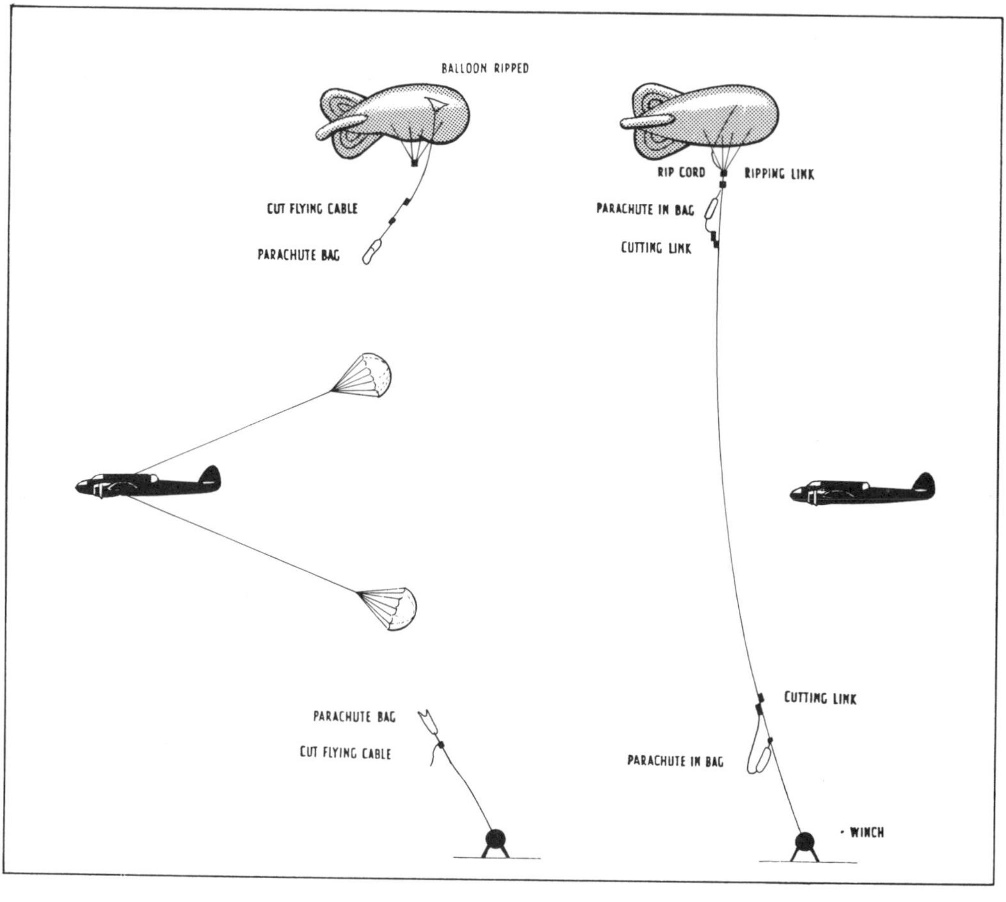

BALLOON RIPPED
CUT FLYING CABLE
PARACHUTE BAG
RIP CORD
RIPPING LINK
PARACHUTE IN BAG
CUTTING LINK
PARACHUTE BAG
CUT FLYING CABLE
CUTTING LINK
PARACHUTE IN BAG
WINCH

acquired meteorological wisdom. Then I looked again, 'That's not a cloud above us,' I thought, 'it looks more like an airship.'

A roar came from Dunn. 'That's a bloody barrage balloon! Where the bloody hell are we?'

I rapidly worked out our dead-reckoning position. 'Must be roughly over Plymouth,' I replied. 'Look out, there's another one ahead!'

Dunn banked steeply and climbed, 'You bloody idiot!' he yelled, shaking the control column in his fury. 'Don't you know that Plymouth is a prohibited zone?'

'Nobody told me,' I replied, by now thoroughly frightened. 'Look out, there's more of them ahead!'

There seemed to be dozens of them as we weaved and climbed. It was utterly terrifying. If we had hit one of those cables, the impact and drag would have sent us crashing to the ground. We dared not turn back and could only trust to our eyesight and our luck.

The fates were with us, for we emerged safely. Another message came through; St Eval was now clear and we were free to land. After we touched down, Dunn was still livid with rage at my ineptitude and I could only protest weakly that I had had insufficient briefing and indeed had never flown at night before. But worse was to follow for the story of our escapade flashed round the squadron and I had to sit squirming with embarrassment and discomfiture in the officers' mess the following day whilst my erstwhile friends and comrades-in-arms made cruel jokes about my navigational ability.

'There's a bigger balloon barrage over London, Nesbit!' someone jeered. 'Why don't you try that one for a change?'

I seethed with impotent fury and stumped off to my room to think things out.

CHAPTER THREE

Salmon and Gluckstein

'There is no more exhilarating feeling than being shot at without result.'

WINSTON CHURCHILL

Stung by ridicule, I was also extremely angry at the poor quality of briefing at St Eval, and certainly this was quite inadequate in early 1941. The procedure was for the pilot and navigator to report to the Operations Room where they were told their destination and time of take-off. The navigator then received a note of the tracks and heights they were supposed to fly and he wrote these on the reverse of the first page of his two-sheet navigator's log. There was then a sketchy meteorological report including an estimated wind speed and direction. If we were lucky, we were given a target map, usually a photograph, and then we waited for the time of take-off.

The preparations was extremely casual but practically all our sorties were solo affairs, and on General Reconnaissance pilots were expected to use their initiative far more than on other squadrons. At the most, only three aircraft per night would fly to the same target, and we rarely received the more detailed briefing of Bomber Command, where the squadron would be mustered in a briefing room and addressed by each expert in turn, Squadron Commander, Intelligence Officer, Meteorological Officer, Squadron Navigation Officer, Gunnery Officer etc.

Determined to overcome these deficiencies, I embarked on a crash programme of my own devising. Managing to find a series of charts which covered the whole of our likely operational area, I cut them up and pasted them together and marked on them every airfield and landmark beacon in south-west England. I haunted the Operations Room and the Photographic Section and acquired a series of target maps of French ports, and spent evenings in darkness in my room switching a shaded torch on and off and trying to imprint the details

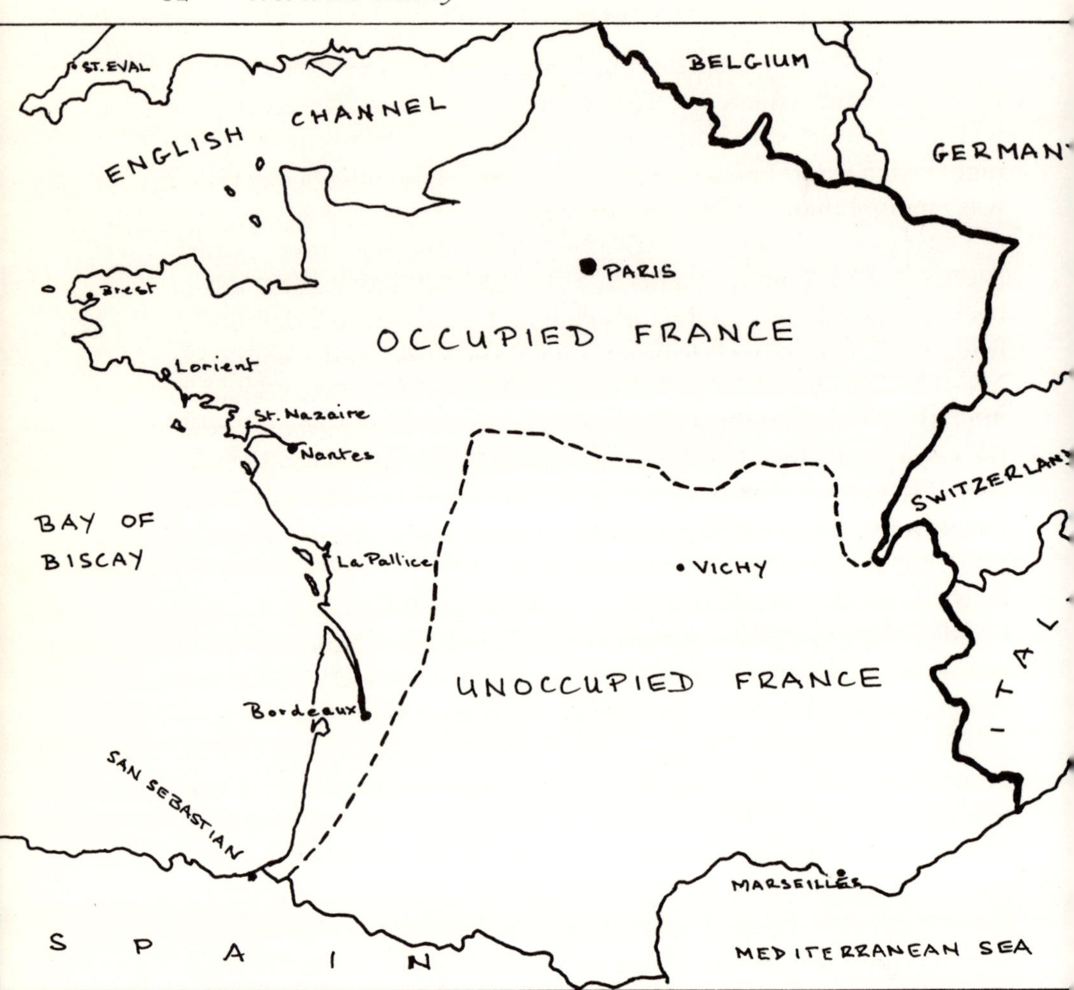

After the fall of France in June 1940, France was divided into two zones. The Northern area was occupied by the Germans but the Southern zone was quasi-independent under the octogenarian Marshal Pétain, who had been one of the most celebrated generals of World War I.

on my mind. I was fortunate in having very good night vision, although I did not realise this until we all underwent tests by looking at a viewer in which the light intensity was steadily diminished whilst a series of ship and aircraft outlines were slid into sight: my night vision was rated as exceptionally good, although my day vision was no more than average for aircrew.

I spent hours practising with a bubble sextant, the Mark 9, which preceded the automatic Mark 9a. Although astro-navigation had not been included in our navigator's course, I taught myself the subject from RAF manuals. I learned star recognition and obtained a Movado astro-watch, which was accurate to within three seconds a day. After some practice I could position myself on the ground using the sextant, the watch and navigation tables, to within about three miles, using observation of three stars; the whole exercise took about ten minutes. But the Beaufort had no astro-dome and there was such refraction from the perspex windows that astro-navigation could not be used, and it was not until after I left the squadron that this skill could be put to good use. A portion of the windscreen panel in front of the pilot was optically true, but the navigator needed all-round vision if he was to obtain his position from sextant readings of three stars or planets. I tried to persuade the engineer officer to devise an optically true perspex cover above the navigator's seat but evidently this proved impossible or more likely was considered unnecessary.

I mentally rehearsed baling out procedures, and also practised moving through the top hatch to the dinghy release in the port wing. During daylight hours I spent hours on the tarmac checking our aircraft compasses for deviation and recording the results on cards for each compass reading. I also studied closely the pilot's cockpit drill until I knew enough to fly the aircraft in an emergency, for I was determined to try to bring it back or at least make a forced landing if the pilot was killed or wounded.

The question of escape from France if we survived a forced landing or parachute jump occupied a lot of my thinking. In those days, we were issued with a few French francs and some packed rations and were told that the best plan was to make our way to Marseilles and seek a character called 'the English Captain' who would somehoworganise a passage home for us. Marseilles was part of

Vichy France, with Marshal Pétain as the nominal head, and the area was ostensibly not occupied by the Germans. But this well-worn information seemed so suspect that I devised another plan. My French was adequate and I carried a spare mackintosh in my navigator's bag, containing compass and maps. With this I would cover my uniform and beg or steal a bicycle on which to make my way down to the Bordeaux region and thence across the western part of the Pyrenees to Santander and the British Consulate. I even contemplated going further, for rumour had it that the Spanish authorities were hostile to British aircrew and apt to lock them up in rat-infested prisons, and I wondered if I could get all the way to Lisbon, and catch the flying boat service back to England. But all this contingency escape planning was not required during my tour of operations.

My activities on the ground must have attracted some notice, for in spite of my error over Plymouth I was appointed Squadron Photographic Officer and later became Squadron Navigation Officer. Probably these appointments came by a process of elimination, for we lost so many crews through accident or enemy action that I was the only suitable person available. By now, my attitude to operational work was becoming more mature, for I had begun to realise that the main hazard was not the enemy but aircraft accident. At this stage, fear and exhaustion did not fray my nerves so badly and I was determined that my part in the crew should be without reproach, and I hoped that the remainder of my crew would be of a similar high standard.

One of the rather dubious privileges of being a commissioned navigator was that one might be required to accompany the Commanding Officer on his necessarily infrequent operational flights, and my next flight was with Wing Commander Bower, who was an elderly thirty-five or so and far above my status in the squadron. Together we took off at 01.45 hours on 15th May and set course for Granville in the Cotentin peninsula, where we were to bomb the port installations. We flew over Dodman Point on the Cornish coast on a clear moonlight night and some fifty minutes later were exactly on course near Guernsey. On arrival at the French coast, however, a sea mist began to obscure the ground and this became thicker as we flew southwards down the coast to Granville. We were flying at about 1,000 feet and I could still map-read, but Wing Commander Bower suddenly decided to bomb from a higher level and we rose above the mist. The moon shone on the top of the mist so that it became impenetrable, and over the target area I had to report that I could not see the port and thus could not release the bombs.

My pilot was not exactly pleased with this announcement and ordered a course to St Malo, our secondary target, but visibility was even lower in this direction. And so we turned northward until the mist cleared and patrolled along the north French coast looking for a likely target but without success. We returned to base after five and a half hours flying, very tired and near our maximum aircraft endurance, bringing back our bombs. Had I been less conscientious I would have released the bombs into the fog, but this did not occur to me.

Two days later, I flew as navigator for yet another pilot, Pilot Officer Welsh, on a long but uneventful search in daylight over the Bay of Biscay, looking for ships and U-boats. Welsh was another experienced pilot, very popular as a raconteur of rather risqué but amusing stories in the officers' mess; he was a steady and reliable flyer who unfortunately failed to return from a night attack some weeks later.

By now, I had been in 217 Squadron for nearly three months and could claim little credit for any prowess. Next, however, I joined my normal pilot for one of the most dangerous attacks of all, against the docks at Brest, which still harboured the contingent of the German battle fleet. This harbour was the primary target for both Bomber and Coastal Commands at that time, for its destruction was considered vital to success in the Battle of the Atlantic and the maintenance of Britain's lifeline by sea with the rest of the world. It is estimated that during 1941, three-quarters of the tonnage dropped in Europe by the RAF fell on Brest. The town was ringed with hundreds of flak guns of all types, and we used to believe that the Germans put up a continuous protective ring of vertical fire around its perimeter whenever the RAF raided the docks, so that each unfortunate bomber had to pass through at least two segments ofof this deadly curtain. Our assumption was probably wrong, for it seems unlikely that the Germans would have been so profligate with their ammunition without some sort of direction-finding against each aerial target, preferably illumination by searchlights, but perhaps their gunners banged away at the least excuse in the heat of the moment. Certainly the fire power was tremendous, and it was also backed by an effective fighter screen, operating both by day and night.

Bomber Command used to drop their bombs on Brest at night from high level, and so only experienced the larger calibre flak shells, but our Coastal Command Beauforts flew in much lower where we were within the range of guns at all calibres. For us, a daylight attack was absolutely suicidal, as had been demonstrated a few weeks before when we had sent three of our squadron into Brest and all had been shot down, killing all the crews. On this later occasion, we were to attack at night with only two Beauforts, mine being the last of the pair to fly into the target.

We took off at dusk and flew to Land's End before turning towards Ushant, and then ascended to a height of 3000 feet up the now familiar estuary towards the docks, keeping our eyes open warily for night-fighters. Our bomb load was a single land mine of 2000 lb, a massive cylindrical object which was code-named 'magnum' and descended with the aid of a small parachute which helped keep it in an upright position during the descent to ensure that the nose pistol would strike the ground and be driven home to the detonators firing the colossal amount of TNT explosive. The intention was to produce the maximum blast effect on the dock area, for we could not hope to penetrate the 4″ main deck armour of the battle-cruisers with such a weapon.

There was no mistaking the port, for already searchlights were probing the sky and flak was pouring upwards, doubtless directed at our first Beaufort, which was timed to bomb before us, whilst Bomber Command were engaged elsewhere on that night. This was my first experience of intense anti-aircraft fire, and I was too interested and astounded to be frightened. I had a grandstand view from the perspex nose of our aircraft, and saw what seemed to be an almost solid mixture of multi-coloured lights hurtling skywards as

BREST

The port, dockyards and roadstead of Brest were of great strategic importance to the Germans in their attempts to win the Battle of the Atlantic and starve Britain into submission. They built a reinforced concrete shelter for their U-Boats at the Port de Laninon (Rade Abri) and in 1941 protected the area with over 1000 flak guns, more than 250 of them of heavy calibre. Brest was a fearsome place to attack, probably the most dangerous in Europe. On 15th February 1941 217 Squadron sent three of their Beauforts into the area and none returned.

Enemy tracer produces a beautiful pattern on the negative when the camera shutter is left open for a while. It seems impossible that an aircraft could fly through such intense flak and yet it is surprising how often the author's aircraft escaped unscathed.

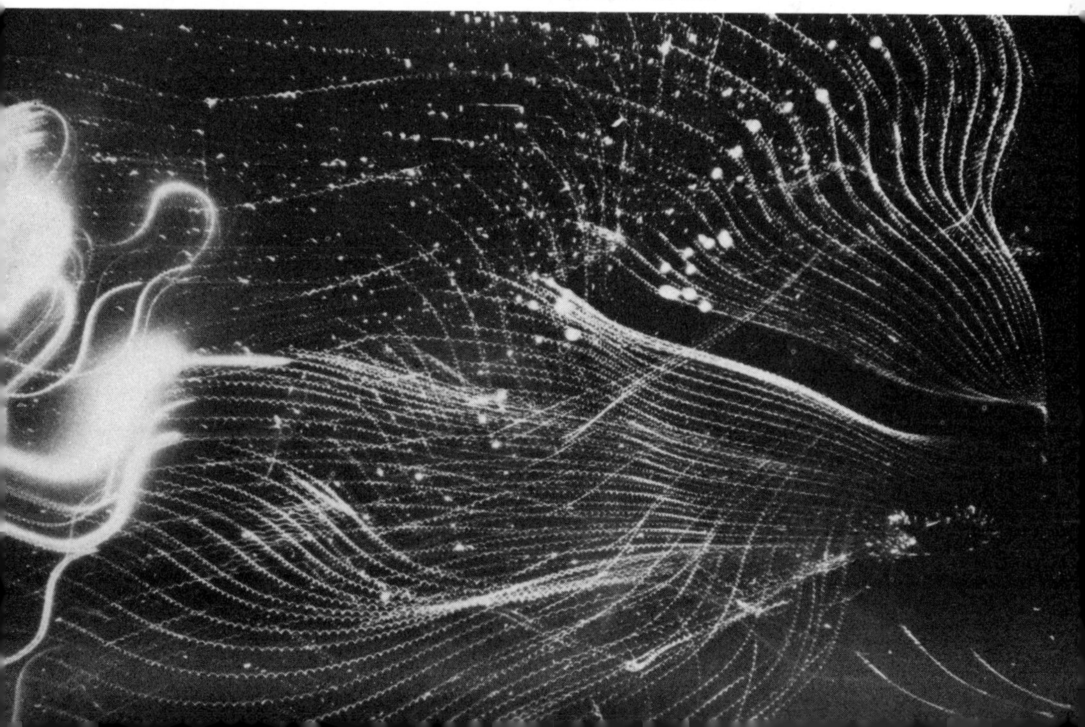

we approached the docks. The Germans used tracer, not only in their machine guns but also in their quick-firing cannons and large calibre flak guns, and they mixed different chemicals in the tracer compound on their shells, which ignited soon after leaving the gun barrels. Thus strontium nitrate could leave a vicious red trail, copper a luminous shade of green, cobalt a rather startling blue, sodium the more normal orange, and potassium a delicate mauve. The utilization of all these different colours enabled each flak gunner to distinguish the trajectory of his own tracer fire and to avoid confusing it with others in the spectrum above him. The Kriegsmarine around the ships had their own colour preferences, whilst the Wehrmacht gunners in the town selected different hues and the Luftwaffe favoured yet another colour combination. Added to all this was the glaring white of the searchlights and the red flashes as the shells exploded. The whole effect was probably one of dramatic beauty, but I was not in the mood for an aesthetic experience in our approaching Beaufort.

We could see this remarkable display clearly but could hear nothing through our helmets, with earphones protected by hemispherical rubber pads, except the over-riding roar of our engines and our tinny voices through the crackling intercom as I gave the pilot his bombing directions and the gunner warned of approaching searchlight beams. We talked with exaggerated calm, as though we were engaged on a training run. My practice on the ground paid off, for I could recognize clearly the enclosure of the Port de Laninon and the surrounding dock area moving down the drift wires of my bombsight. Perhaps the battle-cruisers were also firing at us but I could not distinguish them and now believe that they were protected by camouflage nets which merged into the grey of the sea. The buildings around the port were distinct enough, however, and I had them squarely in the bombsight when I pressed the release button and the Beaufort jolted upwards as the heavy load fell away.

When a bomb is dropped it first takes up the forward velocity of the aircraft and trails underneath the bomb bay, but the powerful tug of gravity immediately brings down the nose and the bomb gathers ever-increasing speed as it falls in a parabola until it smashes into the ground beneath. From such a low level the land mine would have taken barely 15 seconds to fall, even with the descent retarded slightly by the parachute, so that the Beaufort would have flown only a little ahead of the point of impact. I cannot remember if the land mine was fitted with a delay action fuse but, if so, it would have been set to only 10 seconds or so. Certainly as we turned away from that inferno an enormous flash came from the docks and our gunner reported that he has seen the explosion right on target. Talking to the navigator of the other Beaufort the following day, he had to report no results from his land mine, which probably fell into the sea. Only two of us bombed that night, and the Germans admitted heavy casualties.

We did not delay our departure for it seemed impossible that an aircraft could fly through such a barrage and not be hit, but apart from a few buffetings and what seemed like rattlings against our metal-skinned Beaufort, we were unharmed. Visibility was very poor on our return journey and we had to home in using wireless telegraphy bearings, but fortunately St Eval could be seen through the low cloud and we landed just three hours after taking off. It was as satisfactory an attack as one would wish and I was feeling content as we taxied to dispersal point, in spite of the awesome experience of the flak.

The day's action was not over, however, for just as I was packing up my navigational equipment in the nose, a stick of bombs fell right across us, causing violent explosions that rocked the aircraft and blasted lumps of earth skywards. For some reason this attack struck me as grossly unfair.

'Bastard must have followed us in,' I yelled, and rushed down the interior of the fuselage, just in time to see the pilot's boots disappearing through the top hatch. The remaining crew had already left and I dived into the rear turret and attempted to swing the guns round at the Heinkel, which could be seen quite plainly turning on our starboard side, but cursed with frustration when I realised that the pilot, quite rightly, switched off the power. Back I rushed up the fuselage but by the time I reached the cockpit the Heinkel had disappeared, leaving me simmering with fury at my slow thinking.

A truck took the gunner, wireless operator and myself back to the

Course setting bombsight
Mark IXc

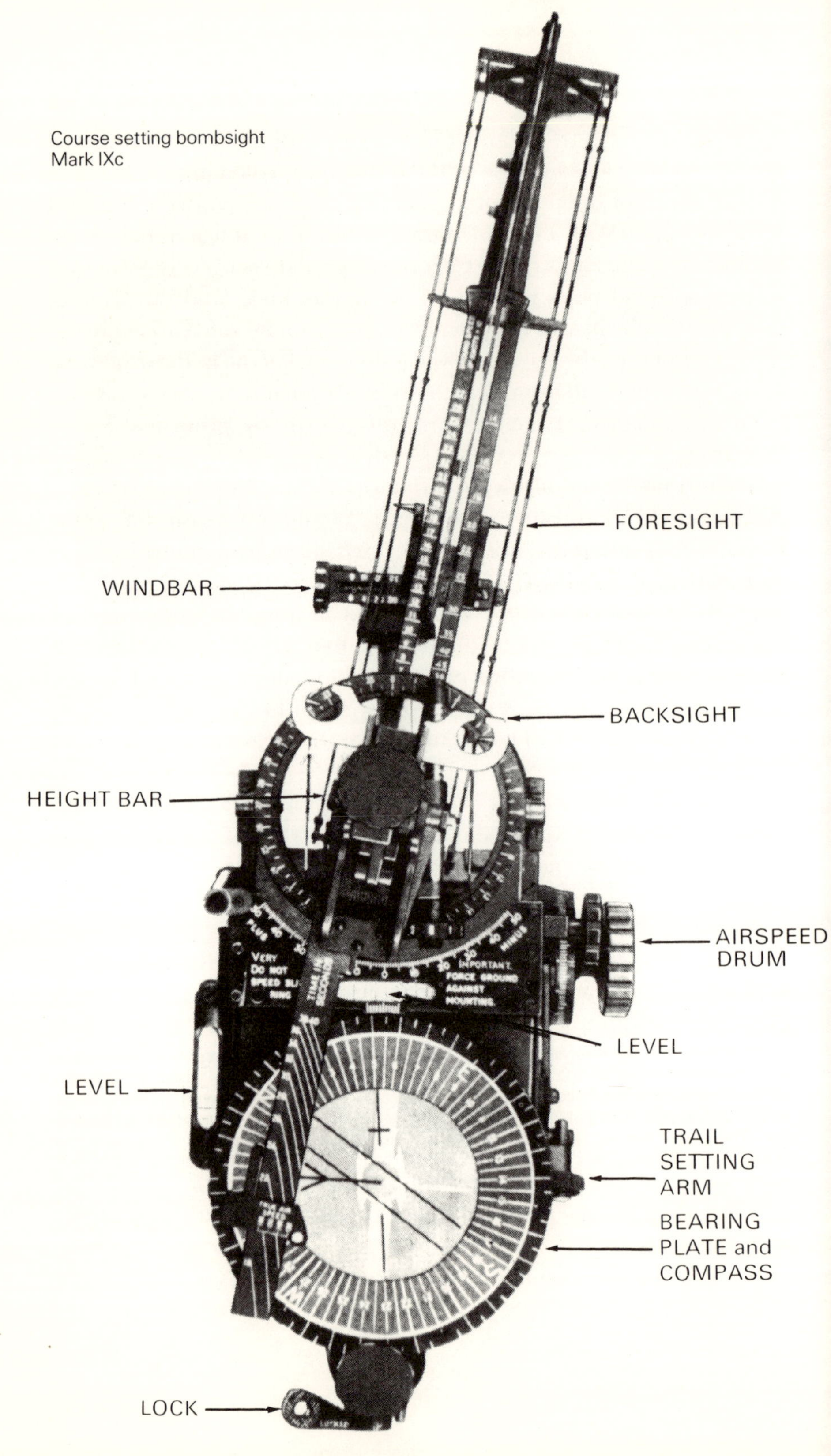

Navigator in nose of Beaufort looking down bombsight.

This instrument may look complicated but it was fairly easy to understand, although much practice was required before a bomb aimer became proficient.

The bomb sight first had to be levelled by turning two levelling screws and bringing the bubbles to their central positions whilst the pilot attempted to fly a dead straight and level course.

Next, the wind speed and direction were set. These were supposed to be calculated over the target but more generally were those that the navigator had worked out en route to the target. The wind speed was set by rotating a drum on the wind speed bar whilst the direction was set and locked on the bearing plate.

The height in feet above the target was set by sliding the backsight up or down the height bar.

The true airspeed was set by rotating the airspeed drum on the right of the bomb sight, which moved the foresight along the drift bar.

The terminal velocity of the bombs was set on the trail setting arm on the right of the bomb sight. This figure was known in advance and would vary according to the type of bomb. A stream-lined armour-piercing bomb fell faster than a semi-armour piercing bomb, which in turn fell faster than a general-purpose bomb, and so on.

In the Beaufort, the navigator was also the bomb aimer. He would instruct the pilot to fly a course which would bring the aircraft over the target, saying 'left-left', 'right', or 'steady', all the time turning the milled head slightly so that the compass dial wires were lined up with the magnetic needle. Meanwhile the aircraft had to fly a steady course through any flak. The target moved down the drift wires and when it appeared lined up with the foresight and backsight, the navigator would press the release button on the automatic bomb distributor and say 'bombs gone' when they had all fallen away.

But at the low level of 700 feet an experienced navigator could possibly dispense with the bomb sight and achieve equally good results by visual observation. Nevertheless, the bomb sight was indispensable as a navigation aid for measuring drift down the drift wires.

Operations Room, which was humming with activity from this sneak raid, but our pilot could not be found and we spent an anxious half-hour before he came in, looking muddy and dishevelled. He had bounded on to the wing of our Beaufort, leaped to the ground, and began running for all he was worth, discarding his parachute and harness en route. Past the perimeter track he had fallen in some mud, picked himself up and continued his headlong flight. Finally, on the outskirts of the aerodrome he paused for breath and found himself alongside a group of curious objects, painted tubes of about four inches in diameter and about three feet high, embedded in the ground. Baffled and curious, he had examined these closely, peering inside the cylinders, and had now arrived to report the mystery. Our Intelligence Officer exploded. These were our latest 'secret weapon' and contained projectiles that shot up in the air and unwound flailing lengths of wire that were supposed to become entangled in attacking aircraft.

'We were just on the point of firing them,' we were told, and I was convulsed with laughter.

'Just as well you didn't sit on one,' I said. 'You'd have really got a rocket!'

By now my pilot was showing signs of strain. He had been flying on operational work for many months and did not like the new Beaufort which he regarded as highly dangerous by comparison with the reliable old Anson. Indeed, the morale of the whole squadron was not particularly high at that time. We had had no remarkable successes with the Beaufort, and there was the over-riding feeling that Britain was alone and that very few of us were doing the fighting. In front of us stretched an apparently interminable conflict, for Germany had not yet invaded Russia, and the forces ranged against us seemed insuperable. Moreover, we did not count our operational flights in the hope of finishing a tour of thirty sorties, as was the practice in Bomber Command. We simply carried on until posted elsewhere by some unseen body in the Air Ministry, or until we were killed, wounded or declared LMF. The latter stood for 'lack of moral fibre' and constituted a thorough disgrace for an RAF officer, especially for a man with a regular commission. Several officers were declared LMF whilst I was flying, but they must remain nameless.

For myself, I was still fairly fresh, and the gnawing fear which destroys confidence and performance had not yet taken hold of me.

For my next operational flight I was detailed to navigate Pilot Officer Kitching to Brest Estuary, armed with a sea mine on this occasion. Sea mines could be of two types, 'floating' or 'ground'. Floating mines were the more familiar variety, moored with a steel cable attached to their anchors on the sea bed and detonating when a ship struck one of the protruding horns; but these mines were invariably laid by surface vessels. The type carried by aircraft was electro-magnetic, and it was dropped at low level in the approaches to a harbour where it rested on the sea bed and was activated by the metal of a ship's hull that passed over it. This mine was a secret piece of equipment, weighing about 2,000 lb and code-named 'cucumber', whilst the operation was called 'gardening'. It was not dissimilar to the German magnetic mine which had caused a great deal of damage to shipping near British ports. The explosive power of this 'cucumber' was enormous for it forced up a terrific pressure, concentrated by the density of the water, tearing the bottom out of a ship. Beauforts were detailed for much of this mining work and apparently the results were quite effective whilst aircraft losses were less heavy than those resulting from low level attacks on port installations.

Kitching and I took off at 03.15 on 29th April and made our way via Land's End and Ushant to the entrance of Brest docks where we dropped this new weapon and encountered less opposition than on my last visit to that infamous locality. None of us liked these early morning sorties, however, for we would be briefed in the Operations Room in the early evening and then told to stand down till about 02.00 hours. Sleep prior to take off, in a small hut on the airfield, was always fitful and we would be awaken by a guard shining a torch in our faces. 'Time to report, sir,' the guard would say quietly and we would stumble into our flying suits ready to be taken by truck in the cold to our waiting aircraft. Spirits were always lowest waiting for the W/T message ordering the take off. The throttles were then opened and as the wheels left the runway it did not take a very sensitive person to wonder if he would ever reach the ground alive. Once airborne there was no time to think along these morbid lines and any further nervous reaction would come after landing and usually when getting wearily back to bed.

It was on one of these early morning occasions when I saw the last of my pilot. Briefed once more to make a bombing attack on Brest, we entered our Beaufort and he found that it was unserviceable, and when we were allocated another aircraft there was also something wrong with the engines, and so we could not fly that night. A day or so later, he was posted away from the squadron to become an instructor, and not only was I very sorry when he left but I was also somewhat upset to be without a regular crew once more.

CHAPTER FOUR

Bingo

'It is not enough to aim; you must hit.'

Italian proverb.

Without a pilot my confidence and enthusiasm had declined to a very low ebb in the early part of May. I seemed to be doing nothing very useful on the squadron whilst most of the other navigators had settled in as crew members and were flying regularly. I was 'out in the cold' and also worried in case my navigational ability was in doubt.

But it was not to be long before my chequered career in the squadron settled down into a steadier routine. A new pilot arrived, Flying Officer John Percival, who had been an instructor on Beauforts at the Operational Training Unit at Chivenor, and prior to this had seen plenty of action in Bristol Blenheims during the collapse of France. Percival was a Canadian from Vancouver, a lanky 25-year-old with a sardonic grin and a terse manner. He did not take kindly to British authority, in spite of his short-service commission in the RAF. He was sometimes openly derisive to senior officers, and was respected rather than liked by these august personages. In the air, he was an extremely careful but skilful pilot, and his policy was that everything in the aircraft must be 'done by the book'. No duties should be shirked but no foolhardy risks should be taken either. He had an extremely cutting and pithy turn of phrase laced with North American slang and his attitude to me when we first formed our crew was uncompromising.

'I hope you're not gonna be like my last goddam navigator!' he said, looking at me searchingly.

'What was wrong with him?' I asked nervously.

'Shat his pants every flight. Over Dunkirk he'd shout "jink" every time we saw a Messerschmitt. Jink! I had enough to do without that

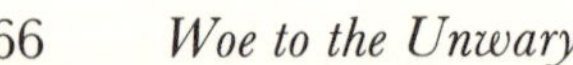

Beauforts leaving Scilly Isles for attack on German surface vessel. This is the actual aircraft in which the author was flying. (*Inset*) John Frederick Percival photographed soon after receiving his short-service commission in 1939.

Mission abandoned, no cloud over French coast. Returning over Land's End.

stupid jerk yelling in my ear,' said Percival.

I assured him that the word 'jink' was not in my vocabulary but that I hoped we would not encounter any Messerschmitts to prove my contention.

To complete our crew we had Sergeant Reeves as rear gunner and Sergeant Davies as wireless operator. They were both extremely competent, and seemed quite unperturbed by our squadron casualties. I had flown with them before and knew their worth. However, the barrier between officers and NCO's in the RAF was fairly rigid at that time and Percival and I did not mix with them socially and rarely saw them between flights. We took them for granted, and they were always there when needed, cheerful, trustworthy and witty, and apparently trusting in both of us. Later in the war this rather absurd distinction between officers and NCO aircrews, many of whom had trained together, tended to break down somewhat.

Our first flight together promised to be exciting. On 13th May we took off to attack an enemy convoy off Guernsey and collected several Spitfire escorts en route in the late afternoon. But we were called back to base after setting course, since either there was insufficient cloud cover or perhaps the report of the convoy had proved erroneous.

The following day we were sent off again, once more in daylight, unescorted on a patrol along the north French coast. This type of patrol became quite frequent in our squadron and was inelegantly called 'Bust' patrol after the more expressive phrase 'Shit or bust'. We took off in the middle of the afternoon and flew along the French coast from Paimpol to Ushant at a height of 500 feet, looking for coastal shipping. There was patchy cloud cover and visibility was good but we saw no shipping other than some small fishing vessels with sails set; they were probably Breton tunny boats, continuing to seek their living in the middle of a ferocious war, and by common assent we always left them alone.

We were still flying at 500 feet with the enemy coast in sight when two loud crashes from underneath the fuselage gave us a severe fright, especially when the aircraft buffeted upwards. At first we thought we had been hit by some mysterious agency but soon realised that our bombs had contrived to fall off, presumably from some electrical fault. The two 500 lb bombs had smashed through the bomb doors and the four 250lb bombs had fallen off the wings. The bombs were not fused. Reeves had seen them splash in the sea, so that the only damage was to the bomb bay and to our nerves. The patrol was almost at an end in any event, but we turned hastily for base and made our report.

A daylight patrol was a hazardous affair and sometimes the Beaufort did not return; we seldom discovered what had happened to missing aircraft. These patrols were always nerve-racking at the best of times, for one could expect enemy fighters to appear, and we would have had little chance of survival unescorted and in broad daylight. There was, however, a certain code of conduct in daylight patrols, for we would not be expected to continue if there was no cloud cover at all and certainly we would abandon the patrol if fighters appeared. We were no match for Messerschmitts and our only hope of escape would be in cloud cover or by extreme low-level flying, at which we were expert. Of course, another purpose of these flights was to check whether the German warships were steaming up the Channel homeward bound from Brest, which in fact occurred about nine months later.

Two days later, on 16th May, we went back to my old haunts in the Brest estuary to drop a sea mine at night, outside the harbour. My navigation log for that night had been lost but we undoubtedly flew at 1,000 feet or so between the indented coasts of that familiar estuary for twelve miles or so, scanning the cliffs and coastline either side of us and knowing that our approach had been signalled ahead and that we would collect our share of flak. Brest had the worst reputation of all the ports attacked by Coastal Command, and was not exactly popular with Bomber Command crews.

Four nights later we were out again and attacked Lanvéoc aerodrome, on the Crozon peninsula, just south of Brest. This was a night fighter station and we identified our target without difficulty, bombing from about 700 feet with 2,000 lbs of bombs armed with delay-action fuses. I could not see any aircraft when releasing the bombs but a number of buildings were plain enough and we directed our load at these. We must have taken the defences by surprise for there was no flak and no opposition, nor did we discover what damage we had done.

By now Percival, Davies, Reeves and myself were working well as a team and had confidence in each other. But Percival was not altogether happy with me. 'Your navigation's fine,' he said, 'but why are you so goddam quiet and calm? You say no more than you have to in the aircraft, and I don't like the way you say "Bombs gone" so casually. We're on bombing attacks, not at goddam Henley Regatta! Pep it up a bit and say "Bingo" instead.'

'Well,' I replied, 'I'm quiet because I'm working hard at my job, concentrating on the navigation and map-reading. And I try to stay calm because that's best for all of us.'

'Sure,' said Percival, 'but from where you sit you can see more than me, and you're the guy with the map. I *need* to know exactly what's going on, and so do the crew. How about giving us a running commentary from the time we hit the French coast? Try to imagine that you're a movie reel commentator.'

'I can see the need for that,' I replied, 'but I'm damned if I'm going to say "Bingo". That's an Americanism and I'm English. But I'll say "Bombs gone" more emphatically in future.'

And so this arrangement was agreed upon and put into practice on our next flight, a low level attack at night on the docks of St Nazaire, on the Loire estuary. This port had already witnessed the tragedy of war, some eleven months earlier. Two weeks after the evacuation of Dunkirk some 90,000 British troops and civilians were gathered in the enclave of St Nazaire, harassed by the advancing Wehrmacht and bombed by the Luftwaffe. The fleeing troops were not a coherent fighting force, and a flotilla of rescue ships awaited them, several of them converted liners. A terrible episode then occurred when about 9,000 men, women and children crammed into the 16,000-ton liner *Lancastria*. Whilst she was at anchor awaiting the tide, four bombs from the Luftwaffe smashed into her so that she sank with the loss of over 3,000 lives; it was probably the worst marine disaster of all time.

Our puny effort on the night of 23rd May 1941 was to attack the port that overlooked this watery graveyard, with four 250lb and two 500 lb of general purpose bombs. We took off at 21.45 hours on 23rd May and flew over the Scilly Islands for a fairly long haul of nearly two hours to the Loire estuary. It may seem extraordinary by modern standards that we should have regarded the navigation of such a flight over the sea as a fairly difficult undertaking, but we had almost no aids. This was in the days before the pulse radio aid of Gee, the radio bombing system of Oboe, and the radar device of H2S which portrayed ground features on a screen, all of which were later installed in British bombers and were to revolutionize navigation and bombing techniques at night. We had to fly at 500 feet above sea level and hope that, after a series of legs around the Armorican peninsula, we would remain undetected beneath the German radar screen and that we would finally arrive exactly at the target area and be able to drop our bombs before the defences were aware of our presence. As yet, we had no automatic pilot, and it was a strain flying the aircraft for so long on steady and accurate compass courses. We could not call for W/T bearings without being heard by the enemy; our standard order was W/T silence once leaving the English coast, except in an emergency, or if navigation bearings were required on approaching base.

During the day it was possible to assist navigation by adjusting the bomb sight so that the crests of the waves below travelled down the wires. Having obtained the 'drift', i.e. the angle between the direction the aircraft was heading and the direction it was actually travelling over the sea, the navigator could next take a compass bearing on the 'wind lanes'. These are narrow parallel furrows in the water which give the direction of the wind and which can be seen by a low-flying aircraft or even from the top of a cliff looking out to sea, if one learns how to recognise them. Putting wind lanes and drift together, the navigator could work out wind speed and direction and arrive at his 'dead-reckoning' position.

At night, there were no aids except careful calculation and concentrated flying, until we made landfall at the French coast and could identify our position on the map. Fortunately the Atlantic surf usually stood out clearly even on the darkest nights, and I could always pinpoint our position fairly quickly and alter course for the target, which at such a low level could usually be identified easily. Any lingering doubt would be removed by the greeting wall of flak at these French ports.

On this night we arrived quite accurately, and the 'running commentary' reconstructed from the navigator's log and from memory, probably ran as follows:

Navigator   'We're about twelve minutes away from the coast, pilot. What height do you reckon we are?'

Pilot   'About 150 feet. Maybe 200.' (The altimeter is set with the day's barometric pressure, which might be a little inaccurate. At this height, the pilot cannot rely solely on his instruments, and always checks visually.)

Navigator   'The ground's pretty low-lying around the target but if we're badly off track it could be up to 500 feet. We don't want to hit a stuffed cloud.' (A stuffed cloud is RAF jargon for cloud-covered hills. The coast is often covered by cloud in the evening, especially on the west coast of France, where warm air moves in from the warmer sea to the cooler land.)

Pilot   'Climbing up to 700 feet. Keep your eyes peeled for night fighters, Reeves.' (Increasing height may bring the aircraft into the German radar screen, so that night fighters would be alerted.)

Air Gunner   'Watching out, sir. Nothing in sight so far.'

Navigator   'ETA at St Nazaire is 23.54 but we should see the coast in about five minutes. I reckon visibility is about six miles.' (ETA means, estimated time of arrival. A mile is a nautical mile, 6080 feet, whereas a normal mile is 5,280 feet.)

Pilot   'I think I see something already. What's that over on the right?'

   (Coastal Command uses a lot of nautical terms, such as port and starboard, beam, bow and quarter, but Percival scorns these and says 'right' instead of starboard beam.)

Navigator   'I see it. Cloud maybe. No, it's a line of surf. But I can't tell where we are for the moment. I was expecting to see land on the port bow.'

Pilot   'Looks like the tip of a peninsula or an island.'

Navigator   'There's more land ahead now, with sea in between these two points. Trying to identify on map.'

   (Shines shaded torch on and off map, alternately peering

Reconnaissance photograph of St Nazaire Harbour. 1) Dockyard Entrance destroyed in the Commando raid of 1942. 2) Minesweepers in outer harbour. 3) U-Boat shelter under construction. 4) Shipyards. 5) French aircraft carrier *Joffre* in dry dock.

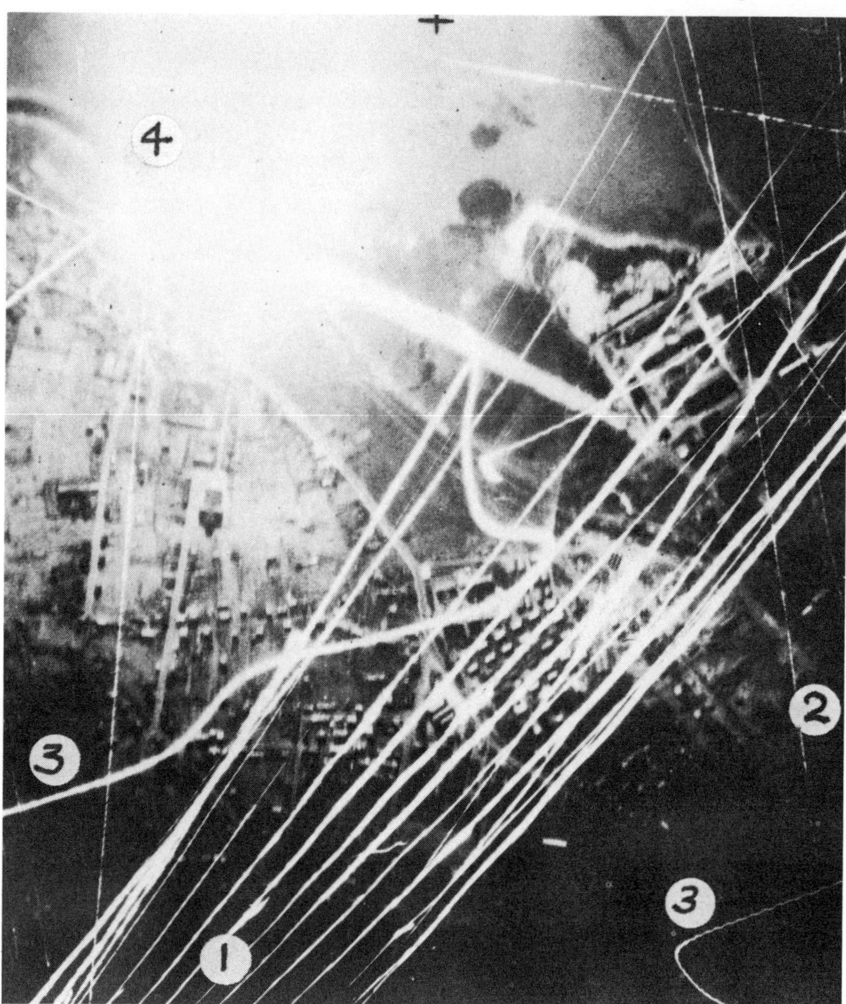

Flak at night over St Nazaire. 1) Light flak. 2) Tracer bullets. 3) Searchlight beams hunting the bomber. 4) Fire in the dock area.

ahead. The eyes become used to quick adjustment to light and dark, with practice.)

Navigator 'I think I've got it. That land to starboard is the tip of an island called Noirmoutier. We're heading towards a peninsula jutting out from the south bank of the Loire estuary. St Nazaire is on the north bank. We're about seven miles south of our track and three minutes early. I'm going to work out a new course.'

Pilot 'Keep that navigator's light dimmed, Nesbit!'
(The navigator's light is an Anglepoise lamp over the chart table in the nose. It might be seen by a night-fighter.)

Navigator 'OK, I've worked it out now. Turn on to 031 degrees magnetic. We should be over the target in six minutes.'
(Magnetic north is not the same as true north. the compass needle points to magnetic north and an adjustment has to be made.)

Pilot 'Turning on to 031 magnetic.'

Navigator 'It's not a bad approach. We're flying towards the Loire estuary, and our ETA works out as planned. We were off course but ahead of time, and the one has cancelled out the other. I can't see the target yet but the banks of the Loire are getting clearer and I can map-read easily.'

Air Gunner 'Searchlights starboard quarter, some distance away.'
(Reeves is sweeping from starboard beam to the tail, then to the port beam and back again. Davies has left his wireless operator's seat and has hopefully set up his Vickers gun in the waist hatch on the port. Both are plugged into the intercommunication and are extremely alert and watchful.)

Navigator 'Those must be from the tip of Noirmoutier. We're travelling away from them. But it may mean we've woken up the opposition.'

Pilot 'How long to target?'

Navigator 'Four minutes. I can see the north bank of the Loire quite clearly now but I can't pick out St Nazaire yet. There are no lights and no flak.'

Pilot 'Good. Let me know as soon as you spot the target.'

Navigator 'Target in sight, commencing run-up.'

(Navigator leaves seat, takes up position over bombsight. He now abandons the nautical terms of port and starboard and uses the standard RAF procedure of left and right for the bombing attack.)

Pilot 'Bomb doors open. Height 700 feet.'

Navigator 'Bombs fused. Bombs selected, 40 foot stick. We're too far to the right. Left, left, steady. About two minutes to go. Docks are quite clear now, still too far to the right. Left, left, steady. Steady, steady, right a bit, steady, steady . . .'

(Tracer curves lazily up from the ground and then whips past the Beaufort. Two searchlights switch on and probe ahead. Bofors fire opens up, and the tracer grows in intensity, but the aircraft flies straight ahead through the flak, without banking or turning.)

Navigator 'Target almost in bombsight. Right, steady. Steady, steady, steady, bombs going, going, going. BOMBS GONE!'

(The aircraft jerks upwards as 2,000 lb of bombs fall away from the wings and the fuselage.)

Navigator 'Turn to port! Flak to starboard! Try 270 degrees.'

(The pilot turns left but says nothing; both hands are required to weave his way through the flak and searchlights, and he has no time to operate his microphone switch. Behind, the bombs explode, fitted with eleven-second delay fuses to avoid blowing up the low-flying aircraft. Nevertheless the force of the concussion can be felt by the crew.)

Navigator 'I think we're getting clear now, Keep flying on 270 for a few moments and we'll turn to Cap Fréhel.'

(Cap Fréhel on the north French coast and is the next turning point; the return track is right over the Armorican peninsula of Brittany.)

Air Gunner 'There's still a searchlight almost on us. Shall I fire at it, sir?'

(Reeves always wants to fire at searchlights. He is sure that if he fired down the beam he could smash the reflector and hit the crew.)

Pilot 'No, Reeves, you'll give our position away.'

(Percival is right, the aircraft is not quite out of danger and at low level someone could easily see Reeve's tracer and bring the flak back towards the Beaufort.)

Navigator   'Turn to 002 degrees magnetic. ETA Cap Fréhel 00.30. Climb to 3,000 feet.'

(This track has been worked out in advance to avoid flying over any known defended areas. There is high ground en route, however, and the pilot must start climbing now. The crew realises that the aircraft will probably be tracked by German radar, but there is not much alternative to flying this way. Returning over the longer sea route would bring the Beaufort to the limit of its endurance.)

Pilot   'Altering course 002. Climbing to 3,000 feet.'

Navigator   'Is she flying OK? That felt like a hit on the starboard wing when we were over the target.'

Pilot   'She's OK, I guess. Can you see any damage from your position, Reeves?'

Air Gunner   'Can't see anything, sir.'

(Reeves has a clear view of the trailing edges of the wings from his turret.)

Pilot   'It was a near miss, I guess. Reeves, what did you see of the bomb explosions?'

Air Gunner   'I could see the bombs hit the buildings around the docks and what looked like a tower. The bombs all exploded OK. I can still see the fires.'

Pilot   'What colour are the fires?'

Air Gunner   'Blue and orange at first, now they are orange and red.'

Pilot   'Keep an eye on them, Reeves, and let the navigator know when they are out of sight, then work out the distance away, Nesbit.'

(This gives Intelligence some idea of the extent of the damage caused, and it is the sort of information that the newspapers like to print.)

Pilot   'Keep your eyes peeled for night fighters, everyone.'

(This is a great danger now. The pilot wants to keep low over the ground but the German radar is tracking the Beaufort and the night fighters from Lanvéoc and Morlaix

When used for torpedo bombing, the Beaufort carried the torpedo slung beneath the open bomb bay. The air tail at the back of the torpedo broke away when it entered the water, and the torpedo armed itself when it travelled through the sea towards the target.

F/O Percival and his crew. From left to right, Sgt Davies (wireless operator), F/O Percival (Captain and pilot), Sgt Reeves (air gunner), P/O Nesbit (navigator). In April 1941, Percival was 25 years old, Nesbit 19 yrs and Reeves and Davies probably 20 years. Average life expectancy for air-crew on the squadron was probably about five months in 1941.

Pilot
may be flying to intercept. No one knows what sort of homing devices they have fitted to their aircraft, which are probably Junkers 88. Once past the French coast, the pilot will bring the Beaufort down to 200 feet or so, and hope to get under the radar screen.)

Pilot 'Davies, how about some coffee?'

(The wireless operator always collects a flask of hot coffee before take off as well as a few biscuits. After the attack, throats are dry and the crew need to swallow something.)

Nobody relaxes as they sip the hot sweet coffee. Everyone is anxiously peering into the darkness. There are still many dangers ahead, not only from night fighters but from engine failure, anti-aircraft fire at the English coast and, above all, from fog or bad weather at base. The crew are buoyed up with the prospect of eggs and bacon after de-briefing, a long sleep and the hope of a day off.

My next sortie with Percival was a return visit to St Nazaire in the late evening of 25th May. The navigation log of this flight is missing from my collection, but we dropped a sea mine at about 23.00 hours outside the harbour, code-named Beach. The trip evokes no special memories and must have been a fairly routine flight. There was far less danger from flak in the approaches to the harbour but more likelihood of an encounter with a night fighter in this longer journey than during our flights to Brest or Lorient, and it seems that we had a quiet night.

On our return from St Nazaire, our day off was denied us. Immediately after this flight, dramatic news flashed around St Eval, as it did around the whole country. We knew that the German battleship *Bismarck* and the heavy cruiser *Prinz Eugen* had left their shelter in the Norwegian port of Bergen on 19th May for an anti-shipping foray. Four days later they had sunk the battle-ship *Hood* with the loss of all but three of her crew and had crippled the battle-cruiser *Prince of Wales*. For us in 217 Squadron, the hair-raising sequel to this exploit was likely to be an order to attack the enemy when the ships came within our range en route to Brest Harbour. On 24th May all crews were put on permanent stand-by.

Some of the plans being discussed as we waited struck me as

The battleship *Bismarck* photographed in Dobric Fjord, Norway, just before she set off with the *Prinz Eugen* on her sortie into the North Atlantic.

suicidal. We were to locate the *Bismarck* with the seven or so Beauforts which were the total we could muster at this time, with myself navigating the CO and leading the whole formation. On reaching the battleship we would circle around 'in Indian file'. At the given signal of a Verey light, we would all dive in simultaneously and drop one of our six semi-armour piercing bombs on the target. The survivors would then resume circling and repeat this procedure a further five times.

I had some idea of the fire power of battleships and thought that, although dive bombing was extremely accurate, none of us could survive a sequence of six attacks on the vessel. My suspicions were even more deep-seated, for I did not believe that 250 lb or 500 lb semi-armour piercing bombs would have been effective against the 5″ main deck armour of the *Bismarck*, or even against the upper decks. In order to penetrate armour plating, it is velocity rather than mass that counts; for instance, an armour-plated bullet propelled at great speed is more likely to punch through armour than a large bomb at slow speed, which would probably break up on impact or bounce off the deck. Even when the force of gravity is combined with the downward velocity of a dive bomber releasing its bombs at low level, insufficient 'Impact velocity' would be built up to have much effect on thick armour plating.

One alternative was to drop our bombs from high level but it was almost impossible to hit a moving target from the 15,000 feet or so that would be required for the bombs to accelerate to a sufficient velocity on impact. At the battle of Midway, the turning point of the war in the Pacific against Japan, all the high level bombers missed their targets, whilst the dive bombers destroyed several Japanese aircraft carriers; but they were dropping on wooden decks, not on thick armour plating.

The most effective attack that our Beauforts could have made was with torpedoes, but it was not until several months later that we received new crews trained in such techniques. Torpedoes were aimed and released by the pilot and required a very high degree of skill, for the aircraft had to be flown straight and level within precise limits of height and speed whilst making allowance for the speed and direction of the target – no easy task when under intense flak or attack by fighters. The Beaufort crews were trained at Abbotsinch, near Paisley in Scotland, where the pilot gained experience on the ground in a modified type of 'Link Trainer' which simulated flying conditions, after which the crews went out over the sea and dropped dummy torpedoes against target ships such as the cross-Channel steamer *Isle of Guernsey*. To ensure maximum chance of hitting a battleship the torpedoes were first set to run deep so as to strike underneath the thicker part of the armour plating, and the aircraft in the squadron had to drop their torpedoes simultaneously from different directions, so that the battleship could not 'comb the tracks', that is take avoiding action which she might do if the torpedoes ran towards her on parallel lines.

Since I did not believe that the authorities had calculated the effectiveness of our dive bombing techniques against battleships, I thought that our proposed attack against the *Bismarck* was a suicidal absurdity and became steadily more morose. We were on stand-by for over two days, not allowed to go back to our billets, and occasionally dozing off in chairs in the mess or in the operations room. As we became more and more tired, dark and unworthy thoughts passed through my mind. Would a convenient but minor navigational error on my part enable us to miss the *Bismarck* completely? No, we might be homed in by the shadowing British warships. Could I make an error by setting the bomb distributor mechanism to salvo instead of single bombs? That was the sort of mistake that was easily made in the heat of the moment!

But in the event, a Fairey Swordfish from the Fleet Air Arm struck the *Bismarck* in the steering gear, with a torpedo, enabling the British capital ships and destroyers to finish her off 500 miles away from the Cornish coast, before she came within our range, and so my cowardice was not put to the test. Probably I would have carried out whatever orders were given to us, but one cannot be sure.

And so a desperately tired squadron dispersed to sleep the clock round, safe for another day or so.

CHAPTER FIVE

Shining like a Ruby

'The worse the passage the more welcome the port.'

Thomas Fuller (1732)

Shortly after the *Bismarck* episode I was given eight days leave, a welcome break for me after the strain of several months flying or stand-by. The train journey back to Paddington in war time was very slow, maybe thirteen hours or so, and usually the train stopped at almost every station. However, we were on the end of the line at Padstow and, since I now qualified for a first class ticket, there were plenty of empty compartments when the train set off. Very tired, I tipped up the arm rests, stretched out in my uniform and went sound asleep. Awakening an hour or so later, I was highly embarrassed to see the opposite seat crammed with civilian passengers whilst more were standing in the corridor. I swung up with mumbled apologies and the standing passengers filed in, smiling at me. It was a small act of kindness which for some reason stays in my memory. The RAF was in very good favour with the public, and it was extraordinary how considerate everyone was towards us. Once, when a group of us tried to enter an overfull lounge bar and a harassed steward tried to prevent us, there was a howl of rage from the civilians inside and the poor fellow was threatened with a punch on the nose. Somehow a table was cleared and four pints of beer appeared as if by magic. Everywhere there was enormous public support for the RAF and the knowledge of this acclaim contributed to our morale and to the continuation of our efforts in those dark days.

At home, there was little to do except rest, and I seemed to need plenty of sleep. Almost all of my friends were away in the services, and living in the house were only my parents and my two younger brothers, both below military age, whereas before the war our house had been a centre for many young people. I had told my parents little of my flying activities for fear of giving them anxiety but they had guessed what was happening. Every day the wireless announced 'Last night, Beauforts of Coastal Command attacked etc. etc.', and I must have been a great worry to my parents, who had already endured one terrible war.

The air raids over our part of London were stepped up as soon as I returned home, and we all spent the first night in our air raid shelter in the garden. The next night, I refused to stir from my bed although the house occasionally shook with the concussions of bombs and eventually I had to give way to the expostulations of my father. Our spaniel dog knew the difference between the warning and the all clear sirens and padded in front of us to the shelter. Once we were all inside, I closed the thick wooden trap door and there was a thwack on the outside. Opening it cautiously, I saw a long sliver of shrapnel standing out from the wood like a spearhead, and I levered it out and kept it for a while as a souvenir.

Returning to St Eval, I found that Percival had been taken off flying for a while. He was married at Oxford on 21st June to a charming and attractive blonde English girl named Peggy Trumble, the daughter of a retired flour miller, and brought his bride back to Cornwall where they found accommodation near St Eval. Whilst Percival was off operational work, Davies, Reeves and I flew on six sorties with other pilots. For most of June, Flight Sergeant Wood was allocated to our crew. Wood was a typical Coastal Command pilot, calm and steady, but more accustomed to flying over the sea for long periods in daylight than flying on night bombing raids, and in the next twelve days we were sent on five daytime sorties. Our first flight was on 4th June, a search in the Bay of Biscay, looking for an enemy motor vessel, from Ushant to the vicinity of Lorient. It was a day of low cloud and thick haze and we saw nothing, nor were we seen, and we made a satisfactory landfall back in Cornwall after an uneventful trip.

On the next trip, three days later, uncompromising instructions were posted in the front page of my log by the Operations Room staff: 'BUST Investigate and report, and attack possible enemy tanker St Eval – Land's End – Point de Perm – Ushant – PHCT 3253 – KRQJ 1545 – KRAX – Dodman – Point de Perm – Dodman.' The true spelling should have been Pointe de Pern; the letters were codes for lines of latitude and longitude. In the event, we turned back after approaching Ushant since there was no cloud cover, these being our instructions since it would have been foolhardy to risk at aircraft so exposed to attack by enemy fighters.

Two days later, we were out again, this time to investigate a U-boat sighting south-west of the Scillies. We flew at 500 feet through changeable weather and in the far distance spotted what we thought might have been a U-boat submerging at the end of a long oil streak. We were carrying depth charges and dropped one on this position; it duly exploded but we saw no visible results after circling for a while so that, uncertain of our original sighting, we wasted no more depth charges, and returned home.

On 14th June we again patrolled to Ushant, on a Bust patrol, but on this occasion there was so much thick cloud and fog that flying conditions became impossible. We returned from our patrol but by now St Eval was so fog-bound that Wood had to make an emergency landing at Perranporth, further down the Cornish coast, which he did with great skill in extremely difficult weather. There were no aids to guide us in and a safe landing depended wholly on the expertise of the pilot.

My last flight with Wood was another patrol in the Bay of Biscay, two days later. Once again, visibility dropped to nil and we had to make our way back to base by instrument flying, fortunately being on track and landing safely. These flights were typical of normal Coastal Command work, requiring much skill on the part of the pilot and his crew, dangerous from enemy action, aircraft failure and weather conditions, but undramatic and unrewarding. One additional purpose of these patrols was to look for enemy aircraft and there was one type that we were ordered to attack under any circumstances. This was the Focke-Wulfe Kondor, a long range bomber which was known as the 'Scourge of the Atlantic'. The Kondor had a wing span of over 100 feet, four engines, and a range of about 2,200 miles.

In early 1941 these formidable bombers had caused havoc in convoys in the Atlantic, sinking numerous merchant ships, especially when working in collaboration with the U-boats. Many of

Focke-Wulf Kondor. These large aircraft caused havoc amongst British shipping in 1940 and in 1941 began to co-operate with U-Boats in attacking convoys. As armament on merchant ships increased the Kondors began to suffer casualties, but meanwhile all Coastal Command aircraft were under orders to attack them if a sighting was made.

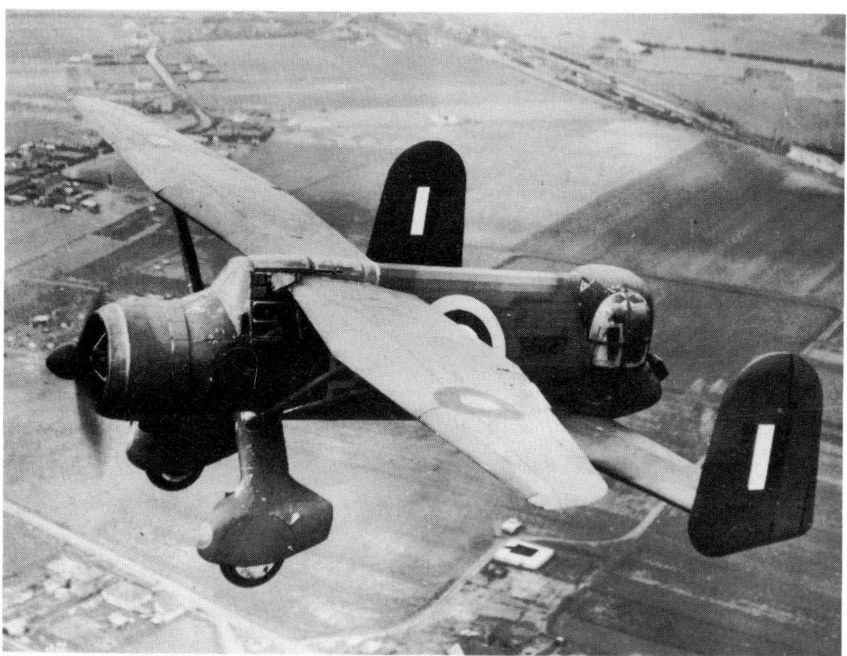

Experimental aircraft at Boscombe Down: the prototype Lysander L6127, with shortened fuselage, tandem wing and four-gun Boulton-Paul turret.

our ships were not equipped with anti-aircraft firepower, and this deficiency had to be remedied to meet the threat of the Kondor. We thought that these raiders were setting out from a base near Bordeaux, ranging into the Atlantic and finally landing in Stavanger in Norway. From time to time, they had been attacked by our Lockheed Hudsons, Sunderlands, Liberators and Catalinas, which had achieved some success although they were slower and less well-armed. The Kondor carried about 4,500 lb of bombs and was armed with 7·9mm and 13mm machine guns, and even a 20mm machine gun. Our ·303 Brownings and Vickers (about 7·5mm) did not compare with these, but the maximum speed of the Beaufort was higher than that of the Kondor, and with our greater manouevreability, we were quite keen to find and attack one of these monsters. But I am not aware of any occasion in which a Beaufort was matched with a Kondor.

On 22nd June yet another pilot was allocated to our crew. This was another Canadian, Sergeant Banning, and we set off on another fairly routine trip, a U-boat hunt in the approaches of Brest estuary. This was again in daylight but once again nothing was seen and again we returned safely. The Atlantic is a very large place and it is surprising how long one can hunt and see nothing in that vast expanse.

Of course, pilots were captains of aircraft irrespective of the rank of their crew. This was quite correct in my view, and when there was an attempt in Bomber Command later in the war to promote navigators to captains, the arrangement proved unsatisfactory. The navigator might do much of the planning and give a flow of directions to the pilot, but the man at the controls had to make the final decisions which would be 'split-second' in an emergency. Doubtless some NCO pilots would have preferred flying with NCO navigators rather than officers, for relations were bound to be a little stilted and formal. In the air, the crew were supposed to address each other on the intercom by their titles, pilot, navigator, wireless operator, gunner etc., and I always observed this code of practice. Percival always used to address his crew by their surnames, however, although I called him 'pilot'. On the ground most officers up to the rank of squadron leader addressed each other by the surnames, or perhaps their nicknames. Percival called me 'Nesbit' and I called him 'Percy'.

Shortly after our flight with Banning, Percival rejoined his crew, only three days after his marriage. I was glad to see him, even though this meant that we would be likely to resume the more dangerous sorties at night. Although he was older than me and one rank my superior, I liked the sardonic Percival and admired his skill; he seemed to me the best man to have at the controls in our hazardous work. He lived outside the station, in a cottage with his wife, so we did not see a great deal of each other when off duty. Neither of us drank more than the occasional half pint of beer, nor was there much drinking at all in the officers' mess and certainly no riotous parties of the type that I experienced later in the RAF. We were very careful not to drink at all if there was any likelihood of flying, but most of us smoked rather heavily, except Percival who never bothered with cigarettes. Percival's idea of enjoyment was to encourage a group of officers to put their arms around each other's shoulders and to sing a rather doleful song called 'Stardust', but the resultant howling noises did not measure up to his more tuneful experiences at high school in Canada.

Our next flight was what for us had become the hazardous routine of flying up Brest estuary to drop our sea mine outside the harbour. We took off on my twentieth birthday, 24th June, one hour before midnight, and as the mine splashed into the water just over one hour later, the Germans wished me Many Happy Returns with a brilliant firework display of tracer and Bofors. Nevertheless it was a reasonably comfortable flight and we were away for only 2 hours 35 minutes, something of a record. Tracer is, of course, continuous fire from machine-guns. Bofors was, and still is, the anti-aircraft gun manufactured by the Swedish firm of that name and used by both sides during World War Two; it was a light gun with a very rapid traverse and elevation, which pumped out a series of explosive shells against low flying aircraft. We assumed that it was this make of gun which the Germans fired at us, but it is more probable that they were using their 20mm flak guns, each of which had four barrels so that the gun could maintain rate of fire of about 750 high explosive shells per minute.

Brest was on the agenda for us again two days later. We set course from Dodman Point to Pierre Noir at the mouth of Brest estuary, and then turned due east up that dangerous alley, dropping our charge and turning on a reciprocal course back to Ushant before heading towards Penzance. My log does not record any flak and we seem to have had a fairly quiet time over enemy territory. Our real problems on this flight did not occur over the target but back in England.

One of our major fears was fog. In retrospect 1941 seems to me to have been a very foggy year in Cornwall. These fogs may have been intensified to some extent by smoke, for most houses used coal for heating, and there may have been some industrial pollution from Plymouth and small factories on the Cornish coast. But the fogs originated in the Atlantic and they rolled stealthily over our aerodrome at unexpected speed. Returning from a raid in the grey early hours of the morning, our nervous energy was usually drained and we would feel stale, flat and miserable, longing to have done with our de-briefing and to return to the relative comfort of our rather sterile rooms and the welcome oblivion of sleep. For myself, I would have preferred to fly through another barrage of flak than roam over Southern England looking fruitlessly for somewhere to land.

We arrived over St Eval at 04.30 hours on the morning of 27th June. Aerodromes were designated by landmark beacons, which flashed the friendly welcome of a morse code signal to returning aircraft, such as Z – X or D – F. We could see the beacon at St Eval when we flew directly over it but a steadily thickening mist had began to obscure the runway so that we could not make a correct approach to land. There were no adequate blind approach facilities, nor did these exist in other aerodromes in SW England, and our safe landing depended wholly on the skill of the pilot.

For forty minutes we circled, attempting unsuccessfully to land. We could see the gooseneck flares tantalising us directly underneath the Beaufort but when circling to make our landing approach we had to peer diagonally through a longer distance of fog, so that the runway disappeared in a woolly blanket. Eventually we were told to circle over Dodman Point on the South Cornish coast and next were instructed to make a landing attempt at Chivenor, Percival's old station in North Devon. At Chivenor, conditions were just as bad as

This unusual photograph
shows a sea-mine, code-
named 'cucumber', in the
bomb-bay of a Beaufort 1. The
sea-mine weighed about
1500lb and was dropped in
sea lanes where it rested on
the sea-bed, being activated
by a ship's hull passing over it.

This Beaufort received an
explosive shell in the port
wing whilst mine-laying. The
crew were fortunate to be
able to land with an
undamaged under-carriage.

at St Eval and after another fruitless attempt to land, Davies brought us further instructions from his W/T, another diversion to Boscombe Down in Wiltshire.

By now, we were not only tired but thoroughly alarmed. Boscombe Down was only about forty minutes flying time away, but we had expended fuel at an excessive rate over the target and with our circling, and seemed to have little more than eighty minutes flying time left. Miserably, I laid off a track for Boscombe and worked out our course. The night was still dark and there was cloud above as well as that baleful fog below. As we flew on our course we could see nothing at all, and when we reached our estimated time of arrival over Boscombe I had to report to Percival that I could not see anything at all through the murk below. We had to decide what to do, for our gauges showed sufficient fuel for only another twenty minutes of safe flying. There was only one answer.

'Prepare to bale out,' said Percival.

We were flying about 2,000 feet above ground level, high enough to bale out safely, but Percival began to climb. The danger of jumping out of an aircraft did not dismay me but my arrival back on earth filled me with foreboding.

'Is my navigation really correct, and are we over a nice flat piece of land like Salisbury Plain?' I wondered.

We had obtained a W/T bearing from Boscombe which seemed to indicate that we were on track, but I had been unable to obtain a positive fix since leaving Chivenor, and the swirling blanket below gave me no encouragement.

'Have we somehow strayed over the Channel?' I thought, for I dislike water and feel panicky when it closes over my head, even though I can swim for fairly long distances.

Like all pilots, Percival wore a parachute strapped behind him and which fitted into his bucket seat in the Beaufort cockpit. The remaining crew wore harnesses with spring hooks on the front on which they could clip their parachute packs; this gave us greater mobility in the fuselage and there was a storage rack for the packs near the entry hatch, but I always stowed mine under the chart table so that I could grab it quickly. Gloomily I clipped my pack into place and tried to remember when the silk folds had last been checked by

FIDO:– Fog Investigation and Dispersal Operation was not in use when the author completed his tour of operations but subsequently saved many returning aircraft from disaster.

the girls in the Parachute Section. I would be first out, by jettisoning the hatch beneath the navigator's position, followed by Davies and Reeves. Percival would point the Beaufort towards the Channel and be the last to leave by jettisoning his side window.

I remembered my mental rehearsals. Yank open the hatch by the emergency handle and let it drop away. Pitch out head first and count a few seconds to get clear of slipstream. Don't panic if you can't find the rip-cord handle. *Bend your head and look for it.* Keep your legs together with plenty of spring in them when you hit the ground. Roll over and grab the parachute lines. I should float down easily – I was only a light weight. But would Percival manage to clear the aircraft safely after we had left?

As we waited for the final instructions I continued to peer morosely through the perspex nose. The weather seemed to be clearing, just a little, and the darkness had lightened.

'Hallo,' I thought, 'surely that's a light ahead, very faint in the distance?'

We were heading almost directly towards it. It was flashing Morse. There it was again! I could just about read it. Dit – dit – dit – dah, dit – dah – dit – dit V – L. Andover! Right near Boscombe. We must have flown right over Boscombe, completely obscured by fog and low cloud.

'Landmark beacon port beam!' I shouted. 'It's Andover! Can you see it, pilot?'

'Sure I see it!' replied Percival, for once demonstrating relief and exultation. 'Shining like a ruby in a nanny-goat's ass!'

'Do you want a course back to Boscombe?' I asked.

'No, now we've found an aerodrome, we'd better try to land,' he replied.

Davies tapped out a message on the W/T to Andover control. But once again the fog and other aircraft frustrated our attempts to land and Percival had to nurse the Beaufort around the circuit for a full forty minutes in the lightening sky before we could line up satisfactorily on the runway lit with gooseneck flares, with an Aldis lamp flashing green at us. Our fuel gauges showed zero as Percival made a perfect landing, and just as we had finished taxying to dispersal point the port engine spluttered and died. Wearily, we explained our predicament to Control and a petrol bowser arrived to replenish our tanks. The fog had cleared by then and Boscombe Down announced that landing was at last possible. A short time later we took off again, and landed at Boscombe within ten minutes.

By now dawn had broken and we could see something of this extraordinary station. Strange aircraft met our eyes, types that we had never seen before and would never reach production. Some of them were hybrid aircraft such as a Lysander with twin tails and a rear gun turret. They seemed somehow grotesque and nightmarish, like mutations dragged from the sea bed or animals on which dreadful experiments had been performed. But perhaps we were feeling slightly lightheaded with strain and tension as we walked to the control tower to receive further instructions. These came at 10.15. St Eval was now clear and we could take off again and fly wearily back to base.

Fog claimed many valuable lives and aircraft up to the end of 1942, and was recognised by all crews as one of their most malignant enemies. Then an astonishing device was installed in several airfields in the Eastern counties. This was called FIDO, the initials standing for Fog Investigation and Dispersal Operation. In essence it was astonishingly simple. Petrol burners were installed along the principal runways of some airfields and these created enough heat to disperse the fog. Moreover, the inviting glow could be seen from afar and the system did not require the lengthy blind approach procedures which cause so much congestion even on modern airfields. It was, of course, enormously extravagant in petrol consumption but it is doubtful if the merchant seamen who risked their lives to bring this valuable cargo across the Atlantic would have been resentful of using petrol for this purpose. I never flew into a FIDO aerodrome but, had it existed in 1941, would have regarded it as an unqualified blessing. The crews who used it described the experience as flying into the gates of Hell, but by 1945 over 4,000 returning bombers had landed safely by means of this miraculous invention.

CHAPTER SIX

Black Magic

'In war the last refinements of science are linked with the cruelties of the Stone Age.'

WINSTON CHURCHILL (1942)

The RAF used to define navigation as 'the art of conducting an aircraft from place to place by observation of terrestrial objects and celestial bodies'. This art was often called 'black magic' by other airmen not versed in the mysteries of navigation. Of course, navigation as practised before the advent of radar consisted of somewhat slow procedures that required a detailed knowledge of dead reckoning on charts, map reading, aircraft instruments, astro-navigation and meteorology. In addition the navigator in the Beaufort was an air gunner, a bomb aimer, a photographer, and a signaller with the Aldis lamp. We used to wear a brevet with an O signifying Observer, whereas those who qualified later wore a brevet with an N signifying solely Navigator, somone who had not always qualified on a bombing and gunnery course.

In fact the navigation employed on our Beaufort squadron was rather rudimentary, simply because there was little opportunity to employ more advanced techniques. When briefed for a target, the navigator would be given certain tracks to follow, such as Base to Land's End to Ushant to La Rochelle, perhaps with some positions in simple code, for example designating the latitude of 48° North as CX and 6° West as GY. The navigator would then draw his tracks on his chart and note down these directions and distances in his log, in advance of take-off. The standard chart used was a Mercator's projection of a scale 1 : 1,000,000 (about 16 statute miles to an inch). In the Mercator's projection the lines of latitude and longitude are drawn as straight lines, even though the longitudes should curve in gradually towards the poles. This distorted the portrayal of the surface of the earth on a flat plane but enabled the navigator to measure direction from the north with his protractor. But of course the distortion increased nearer the poles and this type of chart could not be used above about 75° north.

Having drawn his tracks the navigator next had to work the courses to follow. If there was no wind at all, the true course would be the track, and the true airspeed would be the ground speed, but of course these conditions never prevailed. There was always a wind of some sort, and this would be estimated by the Meteorological Officer. The 'Met men' worked with inadequate information, and their forecasts were far less reliable than those of today which are asssisted by numerous reports, recordings, satellites and computers.

Much of our wartime information was supplied by long-range Wellingtons of Coastal Command, which ranged far out into the Atlantic recording cloud conditions, barometric pressure, temperature etc. It was rather a thankless task requiring great skill on the part of the crew, who had to fly in almost all adverse weather conditions. The 'Met men' would enter all their information on a synoptic chart, and forecast the weather conditions and wind speed and direction for the period of the operational flight. Often their forecasts of the wind were reasonably correct, but low cloud and fog did not seem to be accurately estimated, or perhaps the authorities simply disregarded these major hazards on occasions. The wind direction was always given as the direction from which it was blowing, and the strength in miles per hour, e.g. 310°/30 mph, and the navigator entered this in his log. Next, he had to know the proposed airspeed. This was usually about 140 knots (about 161 mph) with 2,000 lb bomb load, and about 160 knots (about 184 mph) once the bombs had been dropped. The airspeed on the aircraft indicator was not correct, for there was always a calculated instrument error, and also a variation with height and temperature. The navigator had to work out the difference between indicated and true airspeeds, which could be considerable at high altitudes.

The six factors of wind speed and direction, airspeed and direction, and ground speed and direction form what is called a 'triangle of velocities', and knowing any four of these factors, the navigator could work out the other two. This could be done geometrically, by plotting on the Mercator's chart, but also on a

(*Left*) Observer and (*right*) Navigator RAF badges.

Even in the RAF, the authorities did not make a clear distinction between these two badges. At the time the author qualified in January 1941, all navigators wore the Observer's badge, which had been introduced by Army Order No: 327 of September, 1915, and which was understood to be a composite badge for those qualified as navigators, air gunners and bomb aimers. As bomber aircraft became larger and the aircrew functions more complex, navigation became a full-time job. In September 1942, Air Ministry Order No. N 1019/42 introduced the Navigator's badge, with the N and twelve feathers instead of the 0 and fourteen feathers.

The new regulation was retrospective, and only those who had qualified before 3rd. September 1939 were permitted to wear the old Observer's badge. However, most of those who were already wearing the Observer's badge continued to do so, including the author, and refused to switch to the new Navigator's badge.

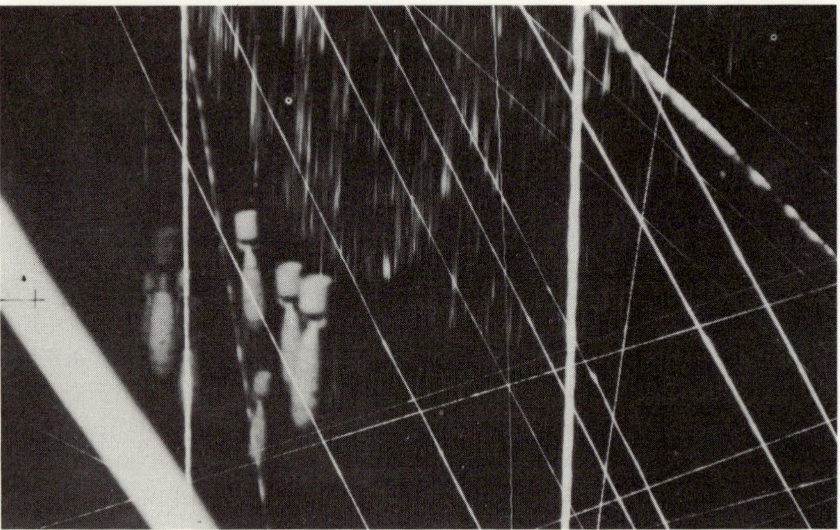

A salvo of bombs has just left the bomb-bay and wings of an aircraft attacking at low level through intense tracer and light calibre flak. In a Beaufort at low level the navigator could often watch the bombs fall all the way to the target, but the bombs did not usually explode until the delay-action fuses were activated eleven seconds later, when the air gunner could see the explosions. The percussions from the explosions usually rocked the Beaufort quite violently.

gadget called the Course Setting Calculator. On this the navigator would set the true airspeed, and then set the estimated wind speed and direction on a dial which he could rotate until the required track appeared against a pointer; another pointer would then show the proposed true course, whilst the ground speed could be read off on an arm which moved in the direction of the track. This was very simple to operate and was really just a mechanical triangle of velocities.

Having obtained his courses and ground speeds, the navigator could then work out the time for each leg of his forthcoming flight. On the back of his Course Setting Calculator was a circular slide rule, calibrated in hours and minutes on the outside scale and distances on the inside scale. Flying time could thus be worked out easily and entered in the log.

The true course is not the same as magnetic course however, for the magnetic needle in a compass is affected by the magnetism in the earth's crust, and the 'variation' which alters slightly every year was shown for all areas on the Mercator's chart. Finally, there was yet another amendment called 'magnetic deviation' caused by magnetism in each individual aircraft, which had to be calibrated for eight headings and stuck on a 'deviation card' in front of the compass.

It would take the navigator maybe half an hour to work out the above calculations, but his advance preparations were not yet complete. He would need to study his maps. These were usually the 1:250,000 series (about four statute miles to the inch) and were 'topographical' in projection, i.e. they represented more realistically the earth's surface with as little distortion as possible. They were coloured and showed heights, rivers, railways, roads etc., just like a large-scale atlas, and the navigator would lightly pencil his proposed tracks on these. Also he usually carried an Admiralty chart which showed the coastline of France in infinite detail. Then there were the target maps or photographs, which the navigator had to study as hard as he could, trying to commit all relevant details to memory, especially when the maps were orientated to the direction in which the attack was to be made.

Lastly, the navigator would record the current call signs and colours of flashing beacons, the recognition colours of the Verey

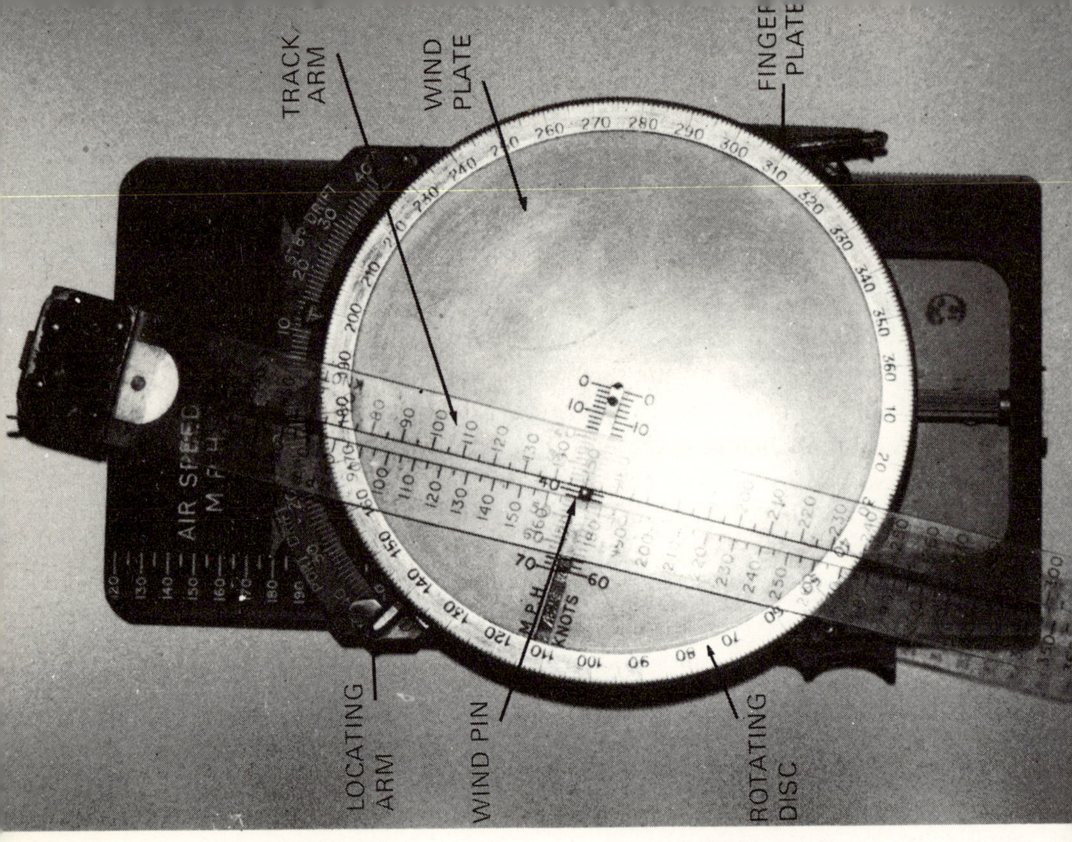

Course and speed Calculator, Mark II.

This calculator was the standard instrument used in 1941 for solution of the 'triangles of velocities'. For example the navigator usually knew, before taking off, the airspeed the wind speed and wind direction, and the direction of the required track, but needed to know the course and groundspeed.

The airspeed was set by depressing the finger plate (12) and sliding along the carriage (4) until the pointer registered against the airspeed scale (150 mph in the photograph.) The wind pin (8) was then depressed and moved outwards until it reached the rotating disc (6). The internal wind plate (7) was then turned until the wind direction was shown on the rotating disc. (158° in the photograph). The wind pin was then moved back until it was registered against the wind speed (50 mph in photograph).

Then the transparent track arm (2) was moved across until its slot was located over the wind pin, and the disc rotated until the required track registered against the moving T pointer.

The true course then appeared against the transparent C pointer, and the ground speed against the wind pin in the track arm.

With a little practice, these calculations could be completed in about 20 seconds.

cartridge for the day and curiously enough, the German recognition colours for the day as well – although we never fired one to test its efficiency.

And so the unfortunate navigator, his mind wonderfully concentrated like the condemned man, would await the time of take-off. Perhaps he would try to snatch a few hours' sleep or cram a few morsels of food into his mouth and an unreceptive stomach. But eventually, an hour before take-off, a truck would take him and the remainder of the crew to dispersal point and the Beaufort. The pilot would usually climb on to the port wing and drop through the top hatch to his bucket seat. The navigator, carrying his navigator's bag and parachute, would enter via a ladder lowered from the port waist hatch, as would the wireless operator and air gunner.

The navigator would enter his compartment and pin his chart with drawing pins on the table, leaving the remainder of his equipment in his bag, since vibration on take-off would shake everything to the floor, and then strap himself into the co-pilot's seat. The aircraft would smell of petrol, oil, dope and metal, and it would usually be very cold and dank. All crew would synchronize their watches with the navigator, who would sit with his log on his knee.

The pilot would go through his cockpit drill and taxi out to the runway for take-off. St Eval was a busy aerodrome and there might be a wait before clearance, after which the aircraft would turn into wind and lumber forward with its bomb load. It was always an anxious few seconds after the wheels left the ground and before we could tell if the Beaufort was climbing smoothly. The pilot would circle to port and the navigator would leave the co-pilot's seat and enter his compartment, unpack his kit and give the pilot his first course, usually on a chit of paper which the pilot would prop above his P.4 compass. The first destination was usually Longships, at Land's End, or Dodman Point, and the next course would be over the sea to France or the Bay of Biscay.

It would usually be extremely cold in the nose of the Beaufort, especially if a blister gun was fitted, for this had an aperture through which ejected shells dropped, and was thus open to the elements. I used to try stuffing my navigation bag into this hole, but it always worked back again. I used to curse this gun, for it always jammed and I never fired it in action, but I would have welcomed a forward firing gun from the navigator's position, and I seem to remember that the enterprising 22 Squadron occasionally fitted a .303 Vickers K for this purpose. The Beaufort was also extremely noisy and vibrated excessively, so that communication was really only possible through the muffled and crackling intercom.

By now, the leaden fear that I felt before take-off would have lifted. The intensity of my work, the need to keep constant vigilance, and the prospect of hitting back at the enemy aroused some sort of fighting instinct in me. Possibly this was a way of expressing resentment at the dread I had just experienced. I would check my calculations again and again, try to find out if we were drifting off course by lining up the bombsight, and keep sweeping the foreground and horizon for signs of anything untoward. I would enter estimated positions at ten minute intervals on the track on the chart, and tick them off as each time passed. I would refresh my mind with the probable appearance of the enemy coastline by looking at the maps yet again. After Percival's strictures, I would keep talking to the others reminding them of the next alteration of course, and when we had to look out for the coast, or other landmarks.

Another function was to keep an eye on weather conditions. Near Brest Estuary, I would usually try to record accounts of types of cloud, temperature, visibility and wind speed and direction, which would be passed on to other aircraft setting out on patrol after we returned to St Eval. Of course, we could recognise the different types of cloud, such as cumulus, stratus, cirrus, etc., unless we were flying in such poor conditions that visibility was nil. On my chart, in addition to the tracks that the aircraft was probably following, I used to keep an 'air plot'. This was a visual record on the chart of the courses in which the aircraft was heading and the distances calculated by the true airspeed. The 'air position' would show where the aircraft would have been if there had been no wind at all; once the aircraft arrived at a known position, such as a point on the coastline, a line drawn between the ground position and air position on the chart would give the wind speed and direction. In fact, I rarely had the opportunity to complete this calculation, for once the French coast came into sight, all my attention was applied to guiding the pilot on the short distance to the target. Entries on charts and logs were all made in pencil. The ball-point pen was not in production until the end of the war, and of course fountain pens would leak with the reduction in barometric pressure at high altitudes.

Many of our flights began in the late afternoon so that as we left the Cornish coast and set off for France the sun would sink and I would make my last checks of the drift of the aircraft through the bombsight before darkness enveloped us. The white caps of the waves would finally disappear, and then the sea would melt into blackness and the horizon would fade in front of us. When Percival had to concentrate on instrument flying at night, I would try to leave my charts and dead reckoning and keep a constant look-out. The navigator's position was excellent in this respect, for I could see ahead, above and below, and even behind us if I craned my neck around. My night vision was keener than the others, and almost invariably I was the first to spot anything, the coast ahead, the faint outline of another aircraft, or the phosphorescent wake of a small vessel beneath us in the sea. I soon learnt to recognise the constellations and knew the names of many stars, even though I could not use the astro-navigation through the curved perspex of the Beaufort, and also I kept track of the positions of the various planets. Once when we were pursued by a night-fighter, as related later, Percival insisted that a planet was a threatening light in our adversary's nose, and would not believe me when I told him that it was Jupiter; but I did not argue, and continued to keep watch on the night-fighter, very faint on our port bow, until it crashed into the Cornish cliffs.

Inside the Beaufort with the navigator's light switched off, the only illumination came from the phosphorescence of the aircraft instruments. Percival's outline showed in a bluey-white light, whilst from behind his armour-plated bulwark came the faint light from Davies' wireless. Behind Davies, Reeves was in complete blackness and freezing cold, with his turret purring as he traversed in a never ceasing watch; but he could not see beneath us, and for our own protection we usually flew very low over the sea. The Beaufort juddered and roared, sometimes plunging and climbing if the air was turbulent. Flying the machine was a terrible strain on the pilot, who had to control the heavy aircraft for up to six hours. If I wanted to talk to Percival other than our brief messages over the intercom, I would leave the nose and sit beside him for a while, taking my map with me; we would lift the earflaps of our helmets and shout at each other above the engine noise.

Percival was six years older than I was, a great difference when one is only nineteen or twenty. He seemed to have confidence in my navigation and map reading, but sometimes intimated that he thought I was stubborn and over-assertive, and he was probaby right. When occasionally other navigators flew with Percival, they found him abrupt and acerbic, but for my part I could not have wished for a better captain, and I liked rather than resented his tough manner, for to me it was a sign of strength. In spite of our differences in character and background, the four of us were probably one of the best crews on the squadron.

The final minutes before the French coast came into sight were always somewhat anxious for me, and for the rest of the crew. We needed to identify our position accurately and as quickly as possible so that we could reach the target area in the minimum of time, in the hope that the defences would not be fully alerted when we made our attack. The weather conditions might be poor, but on the approach from the sea we could usually see the waves from the Atlantic breaking on the shore. But there was very little time to make a positive identification. Flying at 500 feet on a dark, cloudy, misty night, we might only have a visibility of three or four miles, and we would cover that distance in a minute or so. Fortunately the west coast of France is indented, like the Cornish coast, and dotted with offshore islands so that map reading was easier than over a featureless coast. After a while, I learnt to recognise many stretches without the aid of a map. But if we were off track, one had to make a pinpoint as soon as possible, looking out quickly at the ground features and then checking the gap with a dim torch. The secret was to work from the ground to the map, rather than the other way round. Working from the map to the ground, one could inadvertently make the wrong features fit the map, and then the aircraft might be in trouble, heading in the wrong direction, with a confused navigator trying to find out where he had gone wrong whilst the German night fighters could be homing in for an attack with their cannons.

Having made the landfall, a quick alteration of course was usually required and this would mean a few seconds' work on the chart. Then I would switch off the navigation light and concentrate on map reading. I would feel tensed, astonishingly alert, unafraid and quite belligerent. Nowadays, one talks of adrenalin being pumped into the bloodstream, and this is what must have happened. Often, our attacks were solo affairs, so there was a good chance that the Germans might be taken unawares. On other occasions, two or three of us might be engaged on bombing or mine-laying, and if we were not the first to arrive, we could expect a reception committee to be awaiting us.

On the first approach to the target, we would usually be greeted by light flak of some sort, coupled with searchlights. The flak was usually very close to us, but we were a difficult target at night and at low level, and the searchlights only caught us fleetingly or not at all. But the bombing run was very exhilarating indeed especially if we were dropping bombs and could actually feel the force of the explosions and see the resulting fire a few seconds later. Dropping a sea mine was something of an anti-climax, for there was little satisfaction in plopping a large cylinder in the sea outside the harbour mouth; no doubt the Germans went out the following day and swept up many of our mines, but there is also no doubt that this weapon of war was often effective, for both the German battle-cruisers *Scharnhorst* and *Gneisenau* struck RAF mines which caused them a good deal of damage.

The work of navigating on the return trip had largely been prepared in advance. There would have been no time to calculate the first return course whilst over the target, and immediately we were clear I would pass the pre-calculated direction to the pilot. For some miles this course might have to be flown erratically, and I would try to identify another ground position as quickly as possible once we had settled down on a steady course. If we were flying back over the sea route, I could often see our position as we crossed the west coast of France or flew over an island. By the land route to the north French coast, I could rarely identify our position over blacked-out Brittany, and it was difficult to pick out our position as we approached the north French coast from the south. The coast line flying from land to sea does not stand out so clearly as from the seaward direction, and the line of surf might be hidden by cliffs, so that we would be over it suddenly with very little time to map-read.

Then would come the final course back to base, and England was usually foggier than France, with more low-lying cloud. As we neared the Cornish coast, we would often call for W/T bearings. The wireless operator would send out a signal and ground control would take a bearing on this and send us the result, classifying it as a first, second or third class reception. I could probably have managed the navigation without this additional aid, but it was comforting to know that we were on track and the Operations Room were interested to know that we were nearing home. The arrival at the Cornish coast was a time for renewed vigilance. We were always apprehensive at the prospect of another dose of flak, but the over-riding fear, and perhaps the worst experience of all, was fog or low cloud. It was with a sigh of relief that we identified the landmark beacon at St Eval and awaited our turn to land.

Our aids for landing were astonishingly primitive by modern standards. We would have switched on our IFF (Identification Friend or Foe) near the end of our return journey, which would have given ground control at St Eval some warning of our approach. The main method of communication was through the wireless operator using his W/T. The pilot could try to speak to ground control through TR9 (transmitter/receiver) but this instrument behaved so erratically and crackled so badly that the attempt to use it was often abandoned. On reaching St Eval, the navigator or wireless operator would flash the letter of the day with the Aldis lamp and sometimes the pilot would fire the cartridge of the day with his Verey pistol. Near the end of the runway the duty officer was installed somewhat precariously in his caravan, and he would return the aircraft's signal by flashing a green or red with his Aldis lamp, reinforced sometimes with a Verey cartridge of the same colour.

The flarepath was lighted with gooseneck flares, contraptions which looked something like watering cans and contained paraffin and a wick which was ignited or snuffed out manually. On the final approach the pilot could switch on his landing lights if necessary and

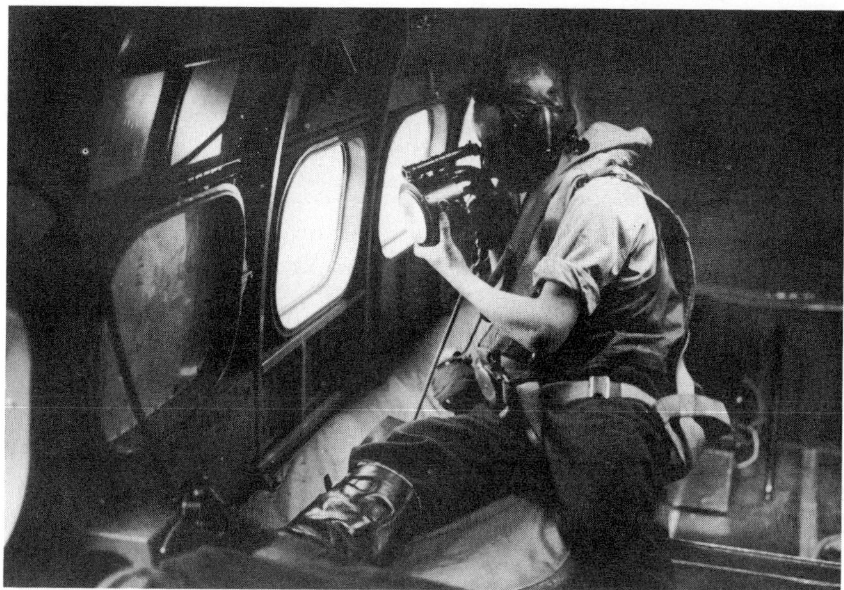

The ubiquitous Aldis lamp was an essential item of equipment in the Beaufort. It could be operated by the wireless operator but more generally was used by the navigator, who had a better field of vision in the perspex nose of the Beaufort. This photograph was taken in a Hudson aircraft.

also there was a 'chance light' at the end of the runway which would sometimes be switched on to flood the runway with additional light if there were no enemy aircraft around. All of these procedures were improved enormously later in the War but we were flying in the most difficult times.

In spite of all the problems, I loved navigating and was in many ways in an element which suited me admirably. I was quick, reasonably accurate, quite fit physically, and eager to improve my prowess. Later in the war, I completed a 'long range navigation course', taking first place amongst thirty or so experienced navigators and obtaining a civilian navigation licence, which I never used. At the end of the war I navigated in the Far East, primarily using astro-navigation as the main method, and it was indeed satisfying to be able to reach objectives over the inky blackness of south-east Asia at night using only star and moon readings from my sextant.

Not all our flying consisted of fear and discomfort in 217 Squadron. The twilights and dawns as we flew could be breath-taking. Once, as we flew back on a clear night over northern France, a sharp frost had covered the landscape and there was a bright full moon. The whiteness of the landscape shone in the clear air, and even I, not renowned for my sensitivity to the finer things of life whilst busily engaged on trying to slaughter fellow humans, could tell that the world was still beautiful.

CHAPTER SEVEN

A Rolling Doughnut

'A weapon is an enemy even to its owner.'

TURKISH PROVERB

By now the war had taken a different turn. On 22nd June the Germans had invaded Russia and our spirits rose with the realisation that Britain was no longer fighting a lone and endless battle. But our routine continued as inexorably as before. Almost every day we were briefed for flying although many of these sorties were cancelled owing to adverse weather conditions or some other reason. We did not count up our operational flights, for there seemed no point in doing so and, as the summer wore on and more and more crews were lost, our chances of survival seemed progressively slimmer and the nervous tension became more acute. One might suppose that one would become inured to danger but the reverse is true, for most people exposed indefinitely to such hazards eventually reach a stage of nervous exhaustion that was called 'shell-shock' in 1914-18 and 'battle fatigue' in 1939-45.

Our crew had not yet reached such a low ebb, although I see from my log book that I had already completed some twenty-five operational flights. Percival must have had an even longer record with previous sorties on Blenheims. By now, we were one of the most efficient and longest serving crews on the squadron, and our teamwork in the air was of a high standard.

It was at about this time that a war reporter flew with us. I cannot bring to mind the name of this reporter, but remember that he was a citizen of the USA who had won a Pulitzer Prize for Literature and had been given an honorary commission as a flight lieutenant in the RAF. He approached me one day at the Waterbeach Hotel and said, 'I believe you have had a lot of experience on night operational work, and I want to try to find out something about how these flights affect you.'

My replies did not satisfy the reporter, for I began to talk about the technicalities of flying and bombing, and he broke in, 'That's not really what I want to know. It is the effect on the crew that interests me. How do you feel when you are dropping bombs, and do you worry about the people that you may be killing?'

My response to this was to say, 'Well, if you really want to know what it feels like, you'd better come on a trip with us. There's a spare seat in the Beaufort and you can probably get permission from our CO.'

To my surprise, he *did* get permission and flew with us on two occasions, although I did not record his name in my log book. In the RAF officers who flew were called 'general duty' officers, whilst others were specialists such as medical officers, equipment officers, engineer officers, meteorological officers, etc. The aircrew officers called the others 'penguins', birds that cannot fly, and just as sailors are supposed to be cruel to sharks and to tease penguins, so we tended to take liberties with our penguins who did not normally take to the air. These officers were usually much older and senior in rank than ourselves, and used to smile indulgently in an avuncular fashion at our antics, probably realizing that most of us had only a short span of life in front of us, and in fact we used to call them 'Pop' or 'Uncle' or 'Doc' or some such name. I was something of a practical joker in my youth and here was my own tame penguin to tease, and I duly thought out a scheme.

In 217 Squadron, we had recently had installed the automatic pilot, now so common that it requires no explanation, but in 1941 its introduction proved a great boon in relieving the strain on the pilot and a benefit to the navigator in that the aircraft flew a straighter course that the pilot could achieve. We called it 'George' after the pre-war catch-phrase 'let George do it', the lazy man's response to any suggestion that he should do some work. When we were well on the way to the target with our reporter, I suggested that he should sit in the navigator's seat in the nose for a while and then persuaded Percival to set the automatic pilot and to slip out of his seat and hide near our lavatory, which was an Elsan pan near the waist hatch.

I then tapped our reporter on the shoulder and said, with great agitation, 'Have you seen our pilot? He's disappeared.'

The reporter looked at the empty seat with dismay and astonishment.

'He must be somewhere,' he cried.

'We've looked everywhere,' I replied. 'We think he must have jumped out. Can you fly a Beaufort?'

Just when our passenger's anxiety was reaching fever pitch, Percival's grinning face appeared from behind the bulwark and calm was restored. Our passenger took this leg-pulling in extremely good part, although one wonders what he must have found to write about. I can remember that he seemed quite unperturbed when we received our nightly ration of flak, but perhaps he did not realise just how dangerous it all was.

My next recorded flight with Percival was on 27th June, yet another night trip to Brest, to drop a sea mine outside the harbour, which had the code name 'Jellyfish'.

On 4th July, we were detailed for a new target. This was the French naval port of Lorient, between Brest and St Nazaire, code-named 'Artichoke'. The Germans were making good use of the ports on the west coast of France. In 1941 they were building massive concrete shelters at Brest, Lorient and St Nazaire, and these were proof against our puny bomb loads of 2,000 lb and even the much heavier bombs of Bomber Command. All that we could do was to mine the entrances to these ports and occasionally try to destroy some of the dock installations. Our task was eventually taken over by Bomber Command which not only destroyed the docks but also most of the towns; but the submarine pens remained intact throughout the war and are now used by the French Navy.

We took off for our new target before midnight and flew west of Ushant. Someone was bombing Brest, for we could see fires and flak, even from a height of 1,000 feet and from a distance of twenty miles. Our track took us south of the Armorican peninsula and we turned south-east to Lorient. Just like Brest, the target lay at the end of an estuary and we had to run the gauntlet of flak and searchlights at our height of 1,000 feet. The run-up to the target was shorter than at Brest, but the inevitable tracer and Bofors that greeted us seemed even more concentrated and accurate as we planted our sea mine near the entrance to the harbour. We turned back with some relief and skirted round the peninsula again for our return journey to Penzance and St Eval.

Our considered opinion was that we preferred Brest to Lorient, and the following night our preference was acknowledged by sending us, without the customary day off, to our old target. Again, we evaded the flak, which was fired at us partly from small ships, probably minesweepers in the estuary, and we returned unscathed.

The next trip was a quite different exercise. We were sent on a search for U-boats *at night*. We took off at 22.00 hours on 8th July and set course from Longships at Land's End for a position 140 miles away in the Bay of Biscay. From there we hunted for U-boats on a clear night, flying a series of short legs south of the Armorican peninsula. We saw nothing and returned home at the limit of our endurance. I have no record of whether we carried bombs or depth charges or why we thought there might be U-boats in that area on that night. Perhaps the flight was the result of some intelligence information, and we were just unlucky not to spot anything.

Two days later we were out again, to drop a sea mine at St Nazaire harbour. If we favoured any target at all, it was this one, for the flak defences on the approach to the harbour were far lighter than Brest or Lorient. Of course, there were always the dangers of night fighters and engine failure, and on this occasion there was the not infrequent problem of a radio diversion to Boscombe Down on our return flight. We located Boscombe without difficulty, however, and were grateful to see that a superb lighting system had been installed around it, giving us a brilliant perimeter of lights, and blue fog funnel lights into the runway which was also very well illuminated. This was the revolutionary new 'Drem' lighting system, and must have been the forerunner of our modern airfield lighting.

We managed a few hours sleep at Boscombe and wandered out at breakfast to look at the strange sights in that bewildering place. Near our Beaufort was a great camouflaged monster of an aircraft, of a type that we had never seen before. It looked powerful, ugly and deadly and we inspected it curiously. Its crew were standing nearby and I recognised an old friend, Sergeant Joe Rickard, who had trained with me. Navigators in the RAF considered themselves a far friendlier breed than pilots, who were just the 'drivers'. We were the

	Reqd. True Track	Distance Run	True Course	Mag. Course	OBSERVATION
	090		088	101	DR 4743N 0451W a/c LORIENT.
					G/S 162 dist 56 time 0-20 ETA 2110
					LORIENT. Dummy town on south bank lit up.
					Cucumber dropped in correct area. Traces fire
	270		272	265	LORIENT a/c 4743N 0451W
					TAS 160 G/S 115 time 0-29 ETA 2145
					ILES DE GLENNAN

Roy Nesbit's account of his operation flying is based on his navigation logs, written in pencil in a vibrating Beaufort, but still legible after forty years. Here he identified a dummy town on the south bank of Lorient, lit up. The abreviations are DR= Dead reckoning, G/S=ground speed; ETA=Estimated time of arrival; Cucumber=Magnetic sea mine; a/c=Alter course; TAS=True airspeed.

In the course of World War II over 85% of Lorient was destroyed by wartime bombing, after which the city was almost entirely rebuilt. This beautifully made model is photographed from the hinterland looking towards the Atlantic and shows the River Scorff flowing under three bridges towards the ocean at the top of the photograph.

'brains' or the 'navigators' union' and our camaraderie transcended rank.

'Does that belong to you, and what is it?' I asked.

'It's a new experimental bomber called the Lancaster,' Rickard replied.

'Can I look inside it?' I enquired.

'No, it's on the secret list,' he replied. 'But I can tell you that we can fly at over 20,000 feet at a speed of nearly 250 mph and carry 10 *tons* of bombs to the farthest corner of Germany. And we are steadily improving on our performance. Moreover, we have secret items of navigational equipment that I can't talk about, but which will revolutionize everything that we used when training.'

I was green with envy and walked around this new beast, guessing correctly that here was something that would make efforts in Beaufort almost wholly ineffective by comparison. Joe was rather smug and superior, and also inclined to be philosophical as we discussed our experiences.

'Season of mists and mellow fruitfulness,' he intoned as we looked at the yellow fog swirling around the aerodrome. 'Close bosom-friend of the maturing sun.'

I was in no mood for Keats and told him that he was quoting about two months out of season. But later that morning when the fog cleared and we flew back to St Eval, I was still suffused with jealousy and convinced that I was in the wrong Command. It seemed to us that Coastal was the Cinderella of the Services and that if there were any new developments in aircraft and equipment, preference would always be given to Bomber or Fighter Commands.

After I left the squadron, I realized that the mysterious device in the Lancaster prototype was probably the navigational aid of Gee. This was devised by scientists at the Telecommunication Research Establishment as early as 1938 and had reached the stage of practical development in 1941. It consisted of synchronized radio pulses sent out from three radio stations in Britain, each about a hundred miles from the other two. In the larger bomber aircraft such as the Lancaster, the navigator did not fulfil the additional function of bomb aimer or gunner, and specialized solely in determining the position of the aircraft and giving the correct courses to the pilot. He was equipped with a special radar receiver, which enabled him to see the time difference between the reception of the various pulses and to plot the results on a special Gee map of Europe which had been prepared with grid lines for that purpose. With three radio stations sending out signals, the navigator could ascertain at which point the position lines intersected or formed a small triangle. The result was accurate to within six miles or so when four hundred miles away from the transmitters, and even more accurate when closer to them. The set was extremely easy to operate and the navigator did not need to understand the scientific basis or the physics involved in this astonishing invention. By March 1942 Gee began to be widely used on bomber raids into Germany but inevitably aircraft were shot down and, although the set had a self destruction device, one was captured intact by the Germans so that by August 1942, jamming counter measures were taken and henceforth the aid could only be used outside enemy territory.

Another aid, used by the Mosquitoes of Bomber Command, was the extremely effective bombing device called Oboe. This employed two transmitters, one at Walmer in Kent and the other at Trimingham in Norfolk. The Walmer transmitter sent out a stream of pulses which the aircraft, flying up the Ruhr in an arc with Walmer at the centre of the circle, amplified and returned. Walmer could tell if the aircraft was to right or left of track and gave instructions accordingly. The second station at Trimingham could tell when the aircraft was in position to drop its bombs and radioed instructions to the navigator; these instructions took account of the aircraft's height, ground speed and track, wind velocity and the terminal velocity of the bombs, and in fact operated as a bombsight directed from afar, so that the navigator could drop his bombs 'blind' but with incredible accuracy. The transmitters could only handle one aircraft at a time, and so high-flying Mosquitoes were used, dropping marker flares which the main bomber force would use as guides. This became the famous Pathfinder force, and the Mosquitoes began their work at the end of 1942 and it was not until early 1944 that the Germans managed to obtain an Oboe set from a downed Mosquito and to begin effective jamming counter measures.

Apart from IFF (Identification Friend or Foe), which was not a navigation aid, there was only one radar invention with which we were favoured whilst I was in 217 Squadron. This was called ASV (Anti-Surface Vessel), although I did not train in this new equipment. In the latter part of 1941, several of our Beauforts were sent to Chivenor in North Devon, where their normal Marconi W/T sets were removed and the ASV installed, the crews returning tight-lipped about the use of the equipment, having been threatened with dire penalties if they breathed a word about its purpose. It was in fact similar to the later device of H2S, used in Bomber Command, and contained a screen over which a radar trace revolved, presenting a fairly effective representation of the coastline, built-up areas, and ships or U-boats on the sea. A detachment of our squadron was sent to Manston in Kent to operate at night in the English Channel with this new equipment, and their operations were shrouded in secrecy and not even entered in our Squadron Operational Book.

Back at St Eval, bad weather halted operational flying for several days and it was not until 15th July that we set off again, in a mine-laying trip to Lorient. Again we received our quota of light flak and again returned to make a safe landing. Then, surprisingly, our crew were given a few days leave, and we returned somewhat refreshed towards the end of July. On our return we were sent off on 31st July to Lorient once again, bracing ourselves for the inevitable flak but returning undamaged.

At last we were given a fairly easy trip, a daylight patrol in the Bay of Biscay. We set off on 1st August, a long uneventful flight in the course of which we saw nothing. However, as we were approaching St Eval at about 1900 hours, near the end of our aircraft endurance, Reeves called from his turret.

'Skua approaching starboard beam.'

A mile or two north of St Eval was another aerodrome, St Mirren, which was occupied by the Fleet Air Arm. Their squadrons used our operational facilities at St Eval, but the Navy flew their aircraft from St Mirren, and we had to keep a watch for each other on landing and take-off. Percival altered his course slightly to avoid the Skua, when Reeves spoke again, more urgently.

'That Skua seems to be in an attacking position, starboard quarter.'

Heinkel 111. The RAF squadrons based at St Eval in early 1941 considered that the Heinkel 111 was their main adversary. St Eval was frequently bombed by Heinkels at night and sometimes by day, whilst Beauforts of 217 squadron retaliated by bombing the Luftwaffe bases at Morlaix and Lanvéoc.

On 1st August 1941, the crew of a Blackburn Skua mistook the author's Beaufort for a Heinkel and attacked with machine gun fire. There may have been a superficial resemblance between the two aircraft. Certainly in both of them the navigator sat in an exposed position in the perspex nose.

A squadron of the Skuas was based at St Mirren, a few miles from St Eval.

I craned my neck around. The Skua was there all right, and moreover the gunner in the rear open cockpit was aiming his Lewis gun at us. A burst of tracer curved towards us, missing by about 25 yards.

'Jesus!' shouted Percival, as he began to take evasive action. I clambered up beside him, with my Aldis lamp.

'How about firing a Verey recognition signal?' I said, and began flashing 'RAF' at the Skua. Our reply was another stream of tracer, passing behind our tail.

'Shall I shoot him down, sir?' enquired Reeves hopefully, in a rather lugubrious voice.

'Naw, leave the bastard alone,' replied Percival, as we climbed into a convenient layer of strato-cumulus just above us, pursued by a long burst of orange tracer.

Emerging cautiously about ten minutes later, we saw no sign of our adversary and made our approach to St Eval and landed. We examined our Beaufort, but could see no bullet holes. As usual, we

The second prototype Avro Lancaster, DG595, undergoing tests at Boscombe Down, before its development into the most famous night-bomber of World War II.

reported to the operations room for de-briefing. Sometimes, one had to queue up for this procedure and in front of us was a very excited sub-lieutenant and his gunner. We waited in silence as he gave the account of a daring attack he had just made.

'We were coming in to land at St Mirren, when a Heinkel 111 came out of the clouds just above us. We manoeuvred to attack and a sort of dog-fight developed. The Heinkel fired at us but we pressed home our attack and fired a full drum at him. The Heinkel was hit in the starboard engine but escaped back into the clouds, streaming smoke and oil. I don't think he could have got far.' Percival and I listened in astonishment whilst the intelligence officer questioned this intrepid pilot. Finally Percival broke in.

'I guess we're the crew of that Heinkel of yours. Except that we were in a Beaufort being fired on by a Skua, and your tracer missed us completely.'

The young pilot's jaw dropped and he yammered something incomprehensible.

Percival looked down at him with bored contempt.

'Go take a running fuck at a rolling doughnut,' he said with quiet venom.

CHAPTER EIGHT

Davies is Good for You

Our nerve-racking routine continued even more intensively throughout August. We were briefed for operational flying almost every day but bad weather hampered us in the first ten days of the month, and it was not until the early morning of the 12th that we made our usual mine laying trip to our private flak-alley in Lorient, dropping our 'cucumber' in the harbour mouth. By now the enemy must have expended a few thousand shells on our Beauforts and the squadron was losing crews steadily, but we always returned. Percival and I were becoming something of a phenomenon on the squadron, the indestructible and lucky pair.

Returning from Lorient on this occasion we were able to snatch only a few hours' sleep before we were sent out in the afternoon for an emergency U-boat hunt in the Bay of Biscay. This was a long and exhausting flight, in good clear weather but with intermittent cumulus cloud at 2,500 feet, which afforded us some protection. We were carrying depth charges and searched our area of the Atlantic Ocean diligently but failed to see any of those elusive conning towers. We returned in the evening dog tired from two operational flights in one day.

Probably one could search the Atlantic for months and not be lucky, but there was always the remote chance of success. Once on a similar exercise on our squadron, however, something had gone terribly wrong. A Beaufort had found a submarine on the surface and had attacked with depth charges. The submarine did not dive and the charges exploded around it. The Beaufort then attacked with machine-gun fire until ammunition was exhausted and radioed for further aircraft. The strike was called off and later a British submarine, badly damaged, managed to limp into its home port in Plymouth. Already in difficulties, it had been unable to dive and was forced to cross through a prohibited zone in an attempt to reach home when our crew had spotted it. This was dreadfully distressing to our crew, their excitement changing to dismay at hearing the news, but they could hardly be blamed for this accident of war. It was not easy to distinguish between German and British submarines from the air, but this episode resulted in us all reaching for our recognition charts.

Our next flight was a mine laying trip to St Nazaire on 16th August, completely uneventful with only moderate flak, and we were thankful to be spared the defences of Brest and Lorient.

Two days later we had to fly to Aston Down and back, for some reason that is now lost, and the next operational flight was mine laying at Lorient on 21st August. The real problem with this port was that, if one was to drop the mine correctly it was almost impossible to avoid flying over part of the defences, and on this occasion we made a wide sweep over the southern part of Lorient. As we were circling south of the estuary I spotted a curious sight on this south bank. A whole new town had been built with lighted streets and simulated buildings. It was easy enough to tell that the town was a dummy from 1,000 feet, but the lights might well have confused a high level bomber. I was able to take very careful note of the position before we dropped our 'cucumber' mine in the entrance to the harbour and set course westwards to the Iles de Glennan and homewards via the sea route and Penzance.

De-briefing took place as usual. First the pilot would give his report and then there was a longer and more detailed account from the navigator armed with his log, and finally the gunner and wireless operator would be asked if they had seen anything. One of the curiosities of these sessions was that the Intelligence Officer would always enter in his records '*pilot* dropped his bombs etc.' whereas it was the navigator who performed this task except when dive bombing. Of course the pilot as captain was responsible for everything but he did not normally drop the bombs any more than he operated the radio or fired the rear guns.

Our Station Intelligence Officer was a Flight Lieutenant Selby, an extremely agreeable person who had lost an arm in the Great War and used to amaze us by the way he could drive his car using his knees under the steering wheel, whilst changing gear with his remaining arm. His junior officer was Flying Officer Eddie Shackleton, son of the explorer, whilst by coincidence our Squadron Intelligence Officer was Flying Officer Gould who had been with Shackleton's father on his South Polar expedition.

Shackleton de-briefed us after the flight to Lorient and I reported the sighting of the dummy town, but this was not believed and I became exasperated at the slur on my ability. As it happened, the position was confirmed exactly by French Resistance forces soon afterwards. Years later I met Lord Shackleton and reminded him of the incident, but it had escaped his memory.

Three days later we were out again on a daylight sweep right to the Ile d'Yeu, the island south of St Nazaire where Marshal Pétain was ultimately imprisoned. We detested these unprotected daylight trips to the French coast, but we cruised all the way there and back without seeing any ships, U-boats or aircraft. We were flying at 1,000 feet and visibility was good, with plenty of welcome cloud cover above us.

Two days afterwards we resumed the now well worn path to Lorient, dropping our sea mine at last light and drawing our usual fire from the port. Returning to Penzance we were met with more anti-aircraft fire from our own defences. In the Beaufort, we had equipment called IFF (Identification Friend or Foe) which we switched on as we approached the English coast and this was intended to differentiate us from German aircraft. Probably there was a naval vessel in the harbour with no means of interpreting IFF signals, and the Royal Navy gunners were notoriously light-fingered with their triggers. One of the Beauforts once landed with a large hole in the wing after flying over Penzance, but the Bofors missed us on that night. We fired off a Verey recognition cartridge, but this made no difference to the enthusiastic gunners and, groaning with despair, we made our way back to St Eval.

Our next flight, on 28th August, was probably the most effective flight we made. Our area of operations was spreading further down the French coast and we were to drop, or so we were told, the first magnetic sea mine outside the harbour of La Rochelle and La Pallice. We took off at 19.00 hours and everything went correctly. We arrived over our target precisely on course and ETA after 2½ hours' flying and also precisely at last light. Visibility was extremely good and this was the only occasion on mine laying when I could clearly see the marker buoys. There were two lines of these leading into the harbour mouth, and I was able to direct the pilot along these and to drop our 'cucumber' exactly between them, a few hundred feet from the entrance. Moreover, we received no anti-aircraft fire whatsoever and made a perfect return trip over the sea.

I was not surprised when a report arrived three days later to the effect that a tanker had struck a mine and sunk in exactly this position. Of course, we could not claim credit for this sinking but I was quite sure in my mind, and still am, that we were responsible.

On 30th August we were given the job of convoy protection, a welcome break from our dangerous night flying routine. Apparently the first major convoy since the outbreak of war was to attempt the passage south of Ireland instead of the safer northern journey. We took off at 08.45 and met this impressive convoy an hour later, counting 24 merchantmen and 2 destroyers. We identified ourselves and circled for three hours with maximum vigilance for U-boats and aircraft, feeling quite proud of the people below us. The next Beaufort relieved us and we flew to Carew Cheriton in South Wales for refuelling, where we were on stand-by for three days in case more merchantmen showed up.

Flying back to St Eval on 3rd September, I was given a most unusual and welcome assignment. We had by this time moved from our quarters in the Waterbeach Hotel, a light and airy little place with a superb view from the cliff edge over the sea and a little sandy beach with a natural swimming pool in the rocks. I believe that the hotel is still there under another name. This was our comfortable officers' mess, and here we were to receive a visit from two very young ladies, Mary Churchill and a companion. Mary was, of course, Winston Churchill's youngest daughter and is now Lady Soames. Two officers were selected to escort them, Jock Maclean and myself, probably because we looked the youngest and the girls were little more than schoolgirls.

An oil tanker sinks in the Atlantic from a collision with a mine, probably similar to the tanker reported to have sunk in the approaches to La Pallice. The davits have been swung out and the crew have got away in the lifeboats.

Jock Maclean was my closest friend. We had trained together and survived in the squadron together. He was a few months older than I and was born on the island of Tiree and his family lived in Linlithgow. When I first met Jock his Scots accent was so broad that I could barely understand him, even though my father originated from the Scottish border, but over the months this accent had toned down somewhat. Jock was a highly intelligent person, well-read for his age, and intended to go on to university. He had a broad grin and an easy manner and was very popular with the girls, whom he attracted without seeming to make any effort at all.

We met the two girls in the afternoon and talked to them politely for some time, answering questions about life on the squadron. The conversation was probably a little stilted but the ice was well and truly broken when Mary Churchill's companion said to Jock, 'So you're Scottish. Where is the smallest airfield in the world?'

'I don't know,' replied Jock politely.

'Under a Scotsman's kilt,' said the young lady.

'What do you mean?' said Jock guardedly.

'Two hangars and a Spitfire!' said the young lady triumphantly.

This promised to be an interesting evening, but someone had muddled up my arrangements.

'If you'll excuse me, I've got to go and bomb La Pallice,' I said 'Maybe it'll be safer there!'

I left in great good humour with everyone's good wishes and at 20.20 hours we took off for what was to be a most unusual flight. La Pallice was at the furthest extension of our range, and we made our way via Land's End to Ushant and then to Le Croisic. From here we headed out to sea, and then turned in towards La Pallice. We were on track but running a few minutes late. I guided Percival at 500 feet towards the docks, with the Ile de Ré to our starboard, giving my usual running commentary. The docks came into view down the bombsight but there were no worthwhile ships in the harbour. Further ahead was a more inviting target, some long buildings which looked like warehouses, and I guided the aircraft towards these, and pressed the bomb release button when they appeared exactly in the bombsight. Then a number of things happened in quick succession.

Near the automatic bomb distributor mechanism there was a red warning light, which came on when the bombs were armed and selected, and should have gone out when the contact arm moved across and the whole stick of bombs had been released. After I pressed the button, and kept it pressed, I twisted my head to watch the warning light. It was still on. One of my recollections of the few seconds of all our attacks is that the events seemed to happen in slightly slow motion, as though my own physical and mental activities were working at a higher pace. I *thought* I had felt the bombs leave the aircraft but at a buffeting low level could not be sure. I pressed my microphone switch and yelled:

'Reeves, did the bombs fall off?'

No reply. Reeves suddenly had his hands full. Several searchlights switched on and one of them caught us immediately, illuminating the whole aircraft and giving us the uncomfortable feeling that a burglar must experience when a policeman's torch shines on him. Reeves did not hesitate, for he was about to achieve his heart's desire. He swung his turret round, depressed the guns and let rip a long burst at the searchlight, which immediately went out, just as he had always forecast.

But I was more concerned with our bomb load, and I yanked out my intercom plug and scrambled to the seat beside Percival. In the Beaufort without inter-communication one had to shout at the top of one's voice and I bellowed:

'I'm not sure if the bombs fell off!'

Percival nodded, his eyes glued ahead. He had heard me call Reeves. The aircraft banked steeply to port, turned back into the dock area, and dived. It was the first time we had made a dive bombing attack other than in practice. We did not dive almost vertically like a Stuka, but even at the shallow angle of 20 degrees the docks and port installations rushed to meet us. Hitherto there had been no flak, for we must have taken the defences by surprise at first, but now yellow tracer hosepiped around us and red Bofors opened up. I could see the tracer streaming at us from gun sites around the docks of La Pallice, and there was even a single gun firing at us from the jetty enclosing the harbour. Percival pressed his bomb release gear and pulled out of the dive. At that moment there were two violent metallic crashes underneath us.

The author believes that a magnetic mine which he dropped at the entrance to the harbour at La Pallice, when navigating F/O Percival's Beaufort on the night of 28/29 August 1941, was responsible for sinking a tanker three days later.

'My God, we've bought it,' I thought, more wonderingly than afraid, although I knew we would have no chance of survival if hit at about 100 feet.

But Percival flew our Beaufort over the sea, weaving from side to side to avoid the following tracer, and we were suddenly aware of a wild-eyed Davies behind us yelling, 'Reeves says the bombs fell off first time!' He had seen four flashes, and we had made an unnecessary second attack.

I dived back into the nose, scribbled few entries in my log, picked up my map and guided Percival along the northern coast of the Ile de Ré, as far as the lighthouse on the western tip, and then set course for home.

When we returned, as we did after a good return flight, no fault could be found in the electrical system. We never discovered what caused the two crashes, for the aircraft semed unmarked, but they may have been the percussions of near misses. Perhaps they were caused by the remaining two bombs falling off, for Reeves had seen only four of our six bombs explode. The remaining two may have 'hung up' momentarily and caused the warning light to remain on; but I have no knowledge of electrical matters.

Reqd. True Track	Distance Run	True Course	Mag. Course	OBSERVATION		
				Bombs dropped – 4 Flashes observed. Ht 500		
				Intense light Flak. Believed A/c hit twice.		
295		292	305	BAL s/c 4643N 0255W.		
				TAS 160 G/S 155 dist 61		
				time 0-23	ETA	2353

Here is an extract from Roy Nesbit's navigational log of the bombing attack on La Pallice on 4th September 1941, scribbled while his pilot F/O Percival was taking violent evasive action. Ht = Height; A/c = Aircraft: BAL = Port des Baleines (a lighthouse on Ile de Ré); S/C = set course; G/S = ground speed; TAS = true air speed.

This attack made the front page of the *Daily Mail* a day later, to our amusement. The crew member who obtained the most satisfaction from the attack was Davies. A strong rumour had circulated in the sergeants' mess concerning beer bottles. It was asserted that an empty beer bottle dropped from a height made an ear-splitting and eerie warbling note, terrifying the ground defences. Davies, who always fretted from inaction when we attacked, had made his personal contribution. He had collected a number of empty Guinness bottles and had hurled them one by one from the waist hatch as we were flying over La Pallice.

'Davies is good for you!' he had shouted happily.

I have never experienced nightmares as a result of my RAF days, but I often have a recurring dream, even after nearly forty years. In this Percival and I are flying very slowly at about 1,000 feet in daylight over a enemy port. Juicy targets are spread below us, ships in the harbour, warehouses, factories, railway stations, and there is always a group of large gasometers. We decide to bomb each of these targets in turn, with a single bomb, for there is no anti-aircraft fire. I guide Percival from one target to the next and we drift slowly towards each objective. I press the release button but the bomb does not fall off, or at least does not explode. There is probably a deep explanation somewhere for this dream, but it eludes me, and all that I can say is that the target looks someting like the complex of La Pallice and La Rochelle.

CHAPTER NINE

Glimmers of Hope

'Dictators ride to and fro on tigers from which they dare not dismount.'

TURKISH PROVERB

After this series of successful flights, Percival and I were to experience three failures. On 10th September we were on stand-by when a report arrived to the effect that a U-boat was making its way back to Brest harbour on the surface. We were ordered to locate this vessel in daylight and attack, in company with another Beaufort and with an escort of three Beaufighters. The Beaufighter was, of course, a development that followed after the Beaufort, and became renowned as a night fighter and for the destruction of surface vessels with cannon fire. We were very glad to have their company as we led the formation towards Ushant. We had expected the protection of some cloud cover, however, but as we approached the target area the weather cleared completely so that we were flying in brilliant sunshine. It was not worthwhile risking our aircraft or our necks on such a vague report and, after communication with base, we turned round and made for home.

Our next trip was the only high level attack I ever made, if 10,000 feet can place us in that category. The purpose of an attack at that height was never explained to us, for the Beaufort did not operate well at high level and all our training and experience was in low level, hit and run raids. The target was the dock area of St Nazaire, and we took off at 01.00 hours on 13th September for the familiar run around the Armorican peninsula, flying initially at low level but gradually increasing height to 10,000 feet as we approached the target.

As usual, I could make a landfall on the coast, for the line of the surf stood out through the mist and it was still fairly easy to map-read. But I found it impossible to pick out the docks of St Nazaire with any clarity. We were fired at, but the flak seemed far less intense than was our normal experience at low level, and we flew to and fro for nearly an hour whilst I tried to see the dock accurately through the bombsight. I was very worried in case our bombs hit French civilian targets instead of the military area around the docks, and Percival persevered with me in spite of the heavy flak around us, from German 88 mm guns, which fired high explosive shells at the rate of about fifteen per minute. Eventually we returned to the coast and, making a definite pin-point over Le Croisic, we set course again carefully for St Nazaire. I dropped the bombs with some misgivings since I really could not be sure of the exact position. Later it became evident that most Bomber Command crews at that stage in the war had to bomb with little chance of hitting their targets accurately. We judged this attack as far safer than almost all our normal flights, but far less effective. Later on, Bomber Command had the use of the radar aids of Gee, H2S and Oboe to bomb over their targets, but prior to this I doubt if they achieved much accuracy, especially over targets deep in Germany and without even a coast line to guide them visually. We had spent so long over St Nazaire that we had insufficient fuel left for a return trip over the sea, and we flew direct to the north French coast.

Somewhere near Lorient we received another salvo of flak, but again this seemed puny by comparison with the tracer and Bofors of low-level. Dawn was breaking as we reached the north French coast and we anxiously scanned the horizon for any sign of the Luftwaffe. I noticed that Percival was sitting with the automatic pilot on and with his head in his hands, although when I asked him if he was feeling ill he shook his head. But I knew that he suffered from sinus trouble, a complaint that was almost an occupational hazard amongst aircrew, exacerbated by extremes of heat and cold with nervous strain thrown in for good measure. The height had affected his sinuses and he was probably suffering acute discomfort. We crossed the coast near Cap Fréhel and set course for Dodman Point and home, almost at the end of our fuel.

Our next trip was a strange affair, and even after all these years, the details are absolutely clear in my mind, even though the circumstances prevented me from recording all of them in my log at the time. We were scheduled to attack the docks at Lorient and took

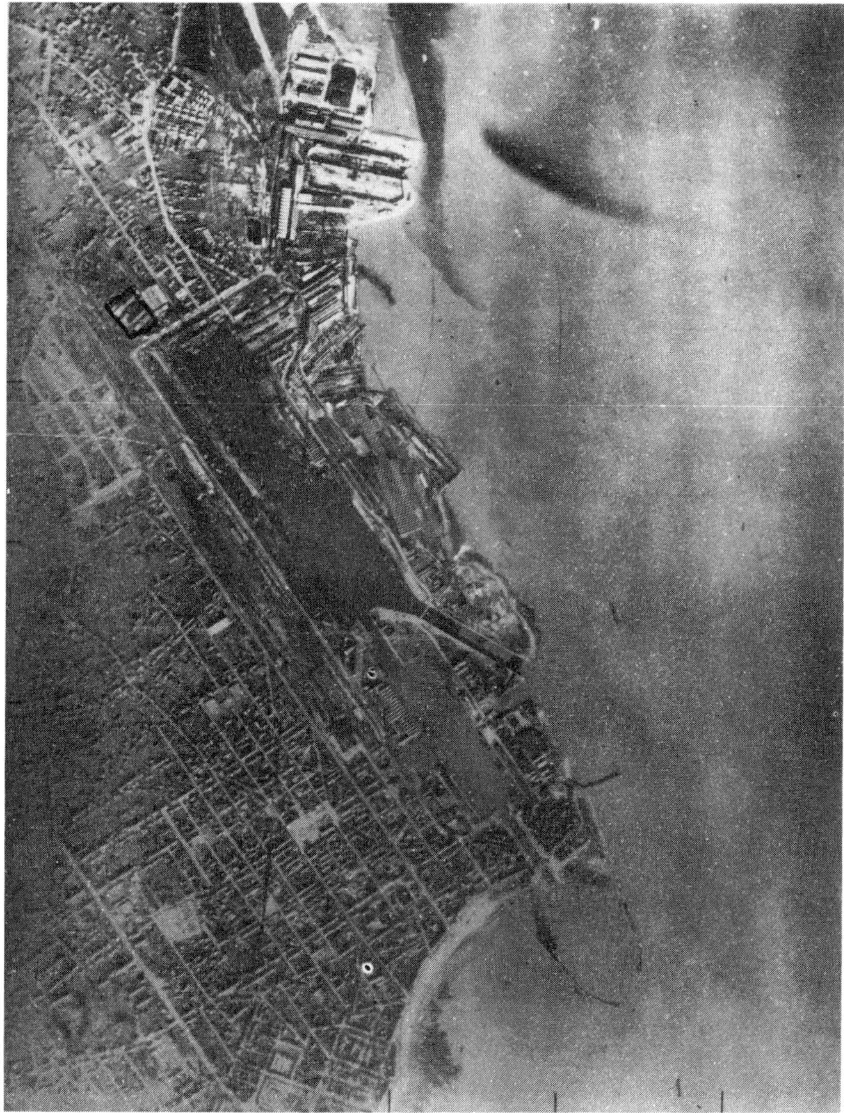

St Nazaire is a major port on the north bank of the Loire estuary. The entrance lock was built in about 1930 to accommodate the great Atlantic liners. In 1941 the German Todt Organisation built the U-Boat base, an immense structure of 960′ × 400′ covered with re-inforced concrete and capable of holding some twenty submarines. British bombers made little impact on this structure, which is now used by the French Navy and is also an industrial site. In 1942 the entrance lock was the target of a determined commando raid.

off at 19.15 hours on 16th September on our usual route. Twenty minutes later we had left Land's End and were well on our way to Ushant when I spotted a strange light, and said rapidly:

'Navigator to pilot. Orange light, port beam, moving starboard. Looks like a night fighter. Travelling under the aircraft, now on starboard bow, moving to starboard quarter!'

As I was speaking, there was a clank as Percival disengaged the automatic pilot and banked the Beaufort to starboard.

'Do you see him, Reeves?' he shouted.

'I see him, sir,' replied Reeves. 'Now on starboard quarter, moving in on us. Shall I open fire?'

'No, Reeves,' replied Percival. 'Do not fire until I say so.'

Percival turned the Beaufort towards the night fighter and dived. The light manoeuvered around us. The shadowy shape of the aircraft had looked like a Junkers 88 as it passed underneath us but we could not be sure of that. And now for ten minutes Percival continued to turn inwards to the light, steadily diving lower. At first, we had been flying at 2,500 feet, higher than normally, for we had been briefed to bomb Lorient from 6,000 feet, but now Percival gradually brought us down to nearly sea level. His reasoning and experience told him that fighters would always try to attack from the rear, under the belly, guided at night by the lights from engine exhausts, and so he continued to turn *towards* the light, diving all the time, whilst we kept spotting the hunter, sometimes Percival, sometimes Reeves and sometimes myself.

At about 200 feet, Percival abandoned our operation, and we turned back to Penzance, with our bomb load, the light still in attendance on our port quarter. One of the items of the navigator's equipment was a coding machine called a Syko, and on this I worked out a message to send to base – 'Chased by night fighter, returning to base, ETA 21.40.' It did not matter about breaking W/T silence, since we had already been spotted. My log is badly kept, for our lights had to be kept dimmed or off, but we crossed the cliffs at Penzance with the orange light still visible, although as we turned for St Eval we could no longer see our menacing adversary.

Landing at base we were met by a buzz of excitement in the Operations Room. They had received our coded message. 'Had the

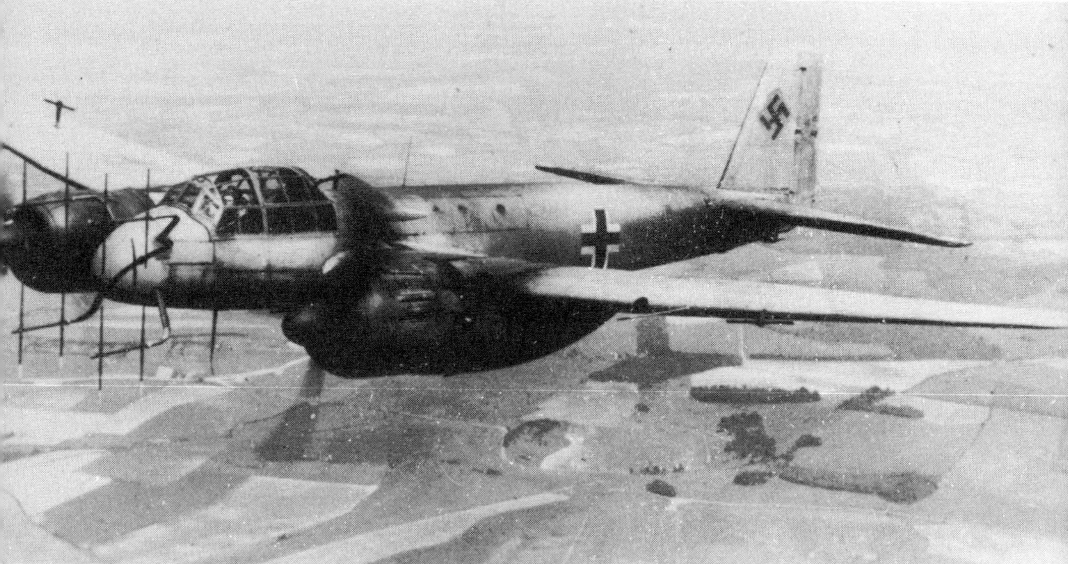

The night fighter that the author encountered in the night of 16/17 September 1941 appeared to be a Junkers 88 with an orange light in the nose, which may have been a searchlight or more probably some sort of homing device. The photograph above shows a Junkers 88 equipped with radar tracking device for night-flying.

Beauforts setting out for a strike, low over the sea.

Donges is an oil port and refining centre. It is situated on the Loire estuary about 11 miles east of St Nazaire. In 1914 the oil tanks of the Consumateurs des Petroles, shown clearly in the photograph, were an important objective for the Beauforts of 217 Squadron.

night fighter fired at us?'

We were not sure, we had seen no tracer.

'Had we fired at the night fighter?'

Definitely not, we did not want to disclose our position.

'Then why did the night fighter crash into the cliffs off Penzance?'

Operations had been following the two blips on the radar; one blip disappeared, and then the report of the crash came in. We had no idea, we said, but could only guess that the pilot was concentrating on his instruments at low level and failed to see the cliffs.

I have never tried to find out why the fighter carried an orange light. It did not appear to be a searchlight and possibly was some sort of homing device, perhaps directed at our engines.

The following day, Percival approached me almost awkwardly.

'I guess you did a good job yesterday, Nesbit,' he said. 'If you hadn't been so alert, we'd have been feeding the fishes by now.'

This was praise indeed from the taciturn Percival, and I could only reply with the truth.

'I was lucky to be flying with someonme who knew how to out-manoeuvre the sod,' I replied, but could not help adding, 'But I hope you noticed that I never once shouted "Jink!"'

Percival could withstand his sinus trouble no longer and was immediately grounded after reporting sick. It might have been a rest cure for me too but some fit of idiocy must have overtaken me. The nights were calm and clear and I had become somewhat fey. I thought I could sense when danger threatened or when the signs of success were propitious, and after a routine flight with Squadron Leader Simmonds, our Squadron Leader Operations, I suggested that he fix us up temporarily with another pilot. Simmonds was an extremely likable and experienced person, who had once returned with his aircraft after being shot through the thigh with flak, and I half hoped that he would fly with us. But his response was to allocate a less experienced but very competent pilot, Sergeant Morgan.

At 18.00 hours on 28th September, Morgan and I took off for a new target, with Davies and Reeves as our remaining crew. This was the oil port of Donges, a refining centre about nine miles east of St Nazaire, on the northern bank of the Loire estuary. Two Beauforts were to make the attack from a height of about 500 feet, and we set off in formation but lost each other in the darkness just before we made an excellent landfall near Le Croisic at the mouth of the Loire. It was a clear starry night above us but sea mist below or perhaps industrial pollution reduced visibility inland, so that I had some difficulty in picking out the target area. But Morgan brought the aircraft down a little to 700 feet and I could identify a leaf-shaped island in the River Loire just south of Donges and we turned northwards for our bombing run. But suddenly two searchlights switched on, both catching us in their beams, and worse followed instantly for the searchlight from a night fighter also picked us out clearly, homing in on our starboard quarter.

Reeves acted without hesitation. He swung his turret round and fired a short burst at one of the searchlights, which went out. It is also impossible to say if he hit either, but his tracers must have had a discouraging effect on both. I could barely hear the gunfire, for the noise did not carry well from dorsal turret to the nose, separated by the roaring engines, and in any event I was concentrating intently on the target ahead. Meanwhile, Morgan did not flinch and responded precisely to my final directions.

Most of my life has been spent in production rather than destruction, but I have to admit that there is something wholly satisfying for a young man of barely twenty to drop a lethal stick of bombs on oil tanks in the face of intense tracer fire and searchlights. We were carrying two 500 lb and three 250 lb of general purpose bombs as well as a canister of incendiary bombs, and a great plume of flames and smoke shot up as these exploded, rocking the aircraft violently. Reeves reported that the flames were reddish brown and we could still see them from a distance of twenty-five miles away before we entered cloud on our return journey over the French mainland.

My navigation log for the return journey is not well kept. We had intended to set course north-east for Les Sept Iles, near Lannion on the north French coast, but we thought we could spot the searchlight of a night fighter constantly apppearing on our port side, and we decided to keep the navigation light off and head due north. Somewhere along the route, flying low over France, we shook off the light and finally crossed the coast near St Mâlo, eventually turning for home from Jersey.

Our next flight with Morgan was to some extent a duplicate of the last one. It was on 30th September, with take-off again around 18.00 hours, but the target this time was Nantes, further up the Loire. We were somewhat off course when on another clear night I identified Belle-Ile, and almost immediately Reeves reported another of those pestiferous night fighters, approaching on our port quarter. We turned immediately and headed out to sea again. My navigation log is inadequately kept from this point but it is still possible to retrace our action. We doubled back on our track, low over the water, banking from side to side to evade our pursuer, for fourteen minutes. By then we seemed to have shaken off the night fighter, and we had seen no sign of cannon fire. After a quick discussion, Morgan turned and we headed once again for Fromentine.

I kept my navigation light off and map read. We flew down the narrow low-lying island of Noirmoutier to the point where a causeway connects the island with the mainland at low tide; nowadays a modern toll bridge and motorway have replaced this ancient passage. From Fromentine we heaed north-east to Nantes, across the flat marshes of La Vendée and southern Brittany. The broad band of the Loire showed the position of Nantes, with the town straddling an island in the middle of the river, but we spent several minutes circling so that I would pick out the target, the Port Maritime on the north bank. Once again, mist was obscuring the target and Morgan had to bring the Beaufort down below the top of the haze so that I could be absolutely sure that I could see our objective. Finally I was able to guide Morgan to the target and we mad a good bombing run, but tracer and Bofors had opened up on us as I released our load, which again consisted of a mixture of general purpose and incendiary bombs. Reeves reported that he saw one of two tall chimneys collapsing as the delay action bombs exploded, and a satisfactorily large fire from our incendiary bombs was visible from over 20 miles away as we headed for Cap Fréhel on the north French coast.

And then something remarkable happened. As we streaked over the French villages and hamlets, I saw lights appearing underneath

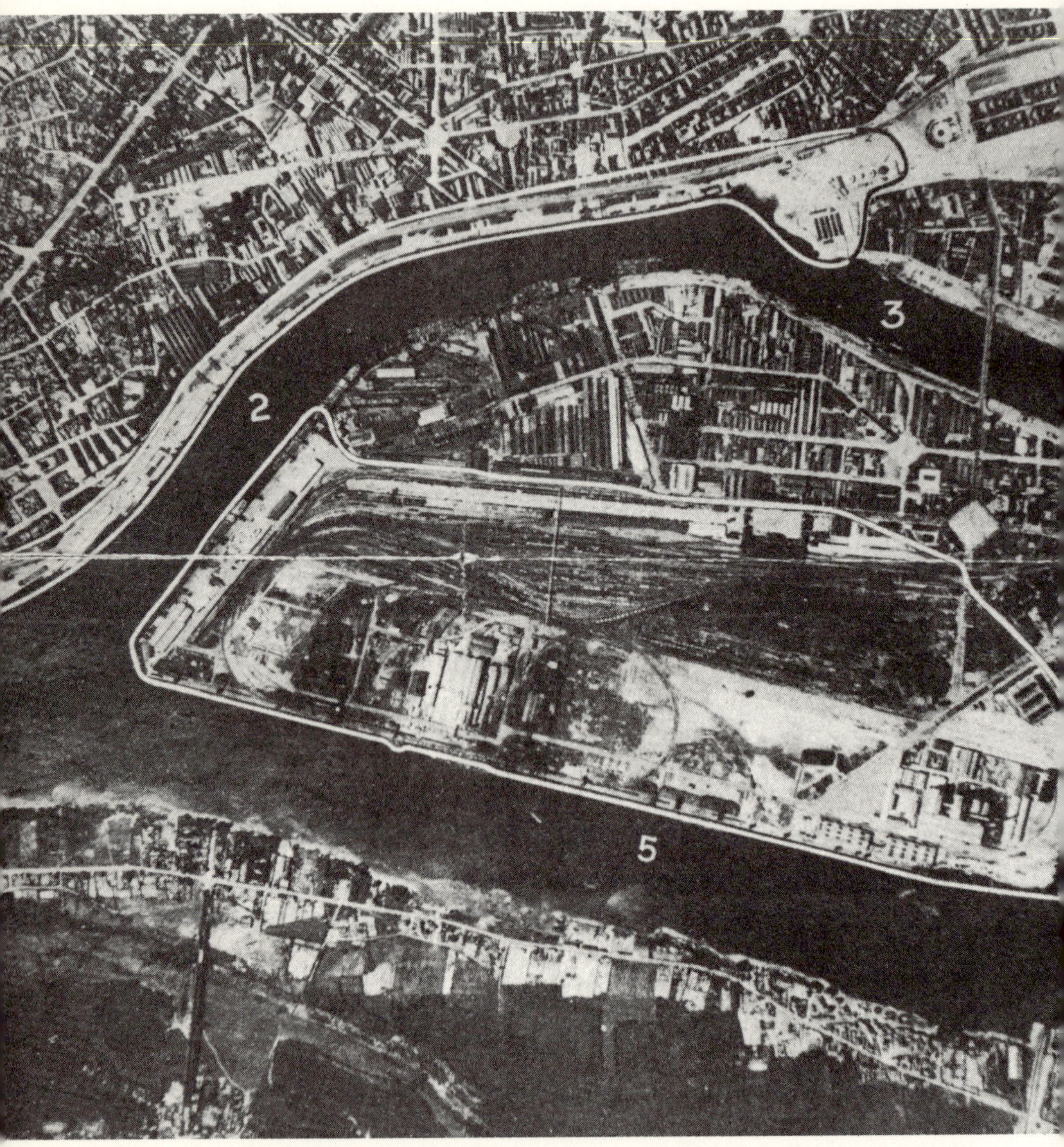

2 = Port Maritime   3= Bras de la Madeleine   5= Bras de Permil
The ancient city of Nantes is the capital of Brittany and straddles the island of Beaulieu in the River Loire. Although Nantes was the sixth most important port in France when war broke out it was not considered such a vital target as Brest, Lorient and St Nazaire, primarily since merchant ships of 20′ draught could only reach the port during high tides. Nevertheless Nantes possessed iron and steel works as well as shipyards, and 217 Squadron was frequently detailed to attack the port. Damage during the war was confined to the city centre and the Loire embankment.

us, coming from what seemed to be houses with open doors. Then there were tiny flashes directed at us. It was Morse. Dit-dit-dit-dah. Dit-dit-dit-dah. We could not believe it at first, but it happened several times. V for victory. Beethoven's fifth. The BBC call sign for broadcasts to Occupied Europe. Those courageous Bretons below were telling us that their spirits were not crushed by the Nazis. Three of our aircraft bombed that night, and we all saw the signals. How hackneyed the V sign seems nowadays, but it was a vital and novel form of inspiration in 1941.

At the time we in 217 Squadron were making our series of attacks on the occupied ports, a strong Resistance movement was growing in France. Within that movement was a small faction of very active Communists, who formed themselves into even smaller mobile 'fireship groups' (*groupes de brûlots*) with the object of striking hard at the Germans by blowing up trains carrying troops or materials, and assassinating German officers, before escaping to another district.

In the latter part of October, a team of three such men, was sent into the town of Nantes. On the morning of 20th October, two of these men, Gilbert Brustlein and Guisco Spartaco, were walking towards the centre of the town looking for an officer to kill. At about 08.15 hours they spotted two German officers walking in the square in front of the cathedral of St Pierre and St Paul and, selecting an officer apiece, they followed them. As the officers reached the pavement the assassins drew their revolvers and fired. Spartaco's revolver jammed but Brustlein fired six shots, hitting his victim twice in the neck. 'He collapsed screaming like a stuck pig,' (*s'effondre en hurlant comme un cochon qu'on égorge*), said Brustlein after the war, for he seemed to have enjoyed his work. The dead Nazi was none other than Lieutenant-Colonel Dr Karl Friedrich Holtz, Feldkommandant of the Nantes Military Region.

The reaction of the Germans was immediate. By the following day they had selected 21 detainees from the prison at Nantes and 27 internees at the camp at Châteaubriant, north of Nantes, and announced that they would be shot 'in expiation of the crime'. The 21 hostages at Nantes, most of whom had been arrested for 'action in favour of the enemy', were taken out and shot immediately. The 21 hostages at Châteaubriant, mostly militant Communists, were loaded into three lorries to be driven to their place of execution; they sang the Marseillaise, echoed by their remaining comrades as they left, and pedestrians uncovered their heads as the lorries passed. Refusing blindfolds, they were shot in three batches at ten minute intervals. A memorial to them may be seen at the town gates of Châteaubriant.

The Germans then announced that a further 50 hostages would be shot if those guilty of the assassination were not arrested by midnight on 23rd October, and they also offered a reward of 15 million francs to anyone denouncing them. A great wave of horror swept around the free world, for as yet people were not inured to such cold-blooded atrocities. Marshal Pétain, head of the Vichy Government, may have been a misguided old man but he was no coward and tried to offer himself as a hostage. The USA, which was not yet in the war, expressed horror and disgust, and Hitler was sometimes sensitive to public opinion in America. The date of the execution was postponed to 29th October.

On the night of 26th October, 217 Squadron returned to Nantes with a force of nine Beauforts, dropping bombs on the dock and scattering thousands of leaflets over the town, displaying on the one side a condemnation by Winston Churchill and a denunciation by President Roosevelt on the other. The following day, Hitler decided to be 'merciful' and cancelled the execution of the fifty hostages, conditional upon the future good behaviour of the people of Nantes. Whilst no one can be sure what prompted this decision, I like to think that maybe our leaflet helped in scaring the tyrant away from further butchery on that occasion.

On 8th November, Sergeant Morgan was awarded a Distinguished Flying Medal for his part in our attacks on 29th and 31st August. He was a remarkably cool and steady pilot who deserved his award; he was later commissioned and fortunately he survived the war. His citation read:

Sergeant Morgan was pilot of an aircraft detailed to attack the oil refineries at Donges one night in September, 1941. On approaching the target, Sergeant Morgan evaded an enemy fighter and enabled his rear gunner to fire a short burst at

Les Otages

DECLARATION DU
Président Roosevelt
SUR LES EXECUTIONS D'OTAGES EN FRANCE

Maison Blanche, Washington
25 octobre **1941**

" La pratique consistant à exécuter en masse d'innocents otages en représailles d'attaques isolées contre des Allemands dans les pays provisoirement placés sous la botte nazie révolte un monde pourtant déjà endurci aux souffrances et aux brutalités.

" Les peuples civilisés ont depuis longtemps adopté le principe qu'aucun homme ne doit être puni pour les actes d'un autre homme. Incapables d'appréhender les personnes ayant pris part à ces attaques, les nazis, selon leurs méthodes caractéristiques, égorgent cinquante ou cent personnes innocentes.

" Ceux qui voudraient " collaborer " avec Hitler, ou qui voudraient chercher à l'apaiser, ne peuvent point ignorer cet effroyable avertissement.

" Les nazis auraient pu apprendre de la dernière guerre l'impossibilité de briser le courage des hommes par la terreur. Au contraire, ils développent leur " lebensraum " et leur " ordre nouveau " en s'enfonçant plus bas qu'ils n'avaient eux-mêmes jamais été dans un abîme de cruauté.

" Ce sont là les actes d'hommes désespérés qui savent au fond de leur cœur qu'ils ne peuvent pas vaincre. Le terrorisme n'apportera jamais la paix en Europe. Il ne fait que semer les germes d'une haine qui, un jour, amènera un terrible châtiment."

Franklin D. Roosevelt

Leaflet dropped by eight Beauforts of 217 Squadron during a bombing attack on Nantes on 26 October 1941. A declaration from Winston Churchill is on the reverse of the leaflet. Note that the declaration from Franklin D. Roosevelt pre-dates the entry of the USA into the war on 6 December 1941.

the enemy and another burst at a searchlight which was extinguished. Finally heavy high explosive bombs and incendiaries were released directly among the oil tanks. All the bombs were seen to burst and their explosion shook the aircraft. A large fire was immediately observed which was still seen to be burning fiercely when the aircraft was some 25 miles from the target. Two nights later this airman participated in an attack on a chemical works at Nantes. On nearing the target area, an enemy fighter was evaded. Sergeant Morgan approached the target and, after overcoming many difficulties, finally succeeded in releasing his stick of bombs on the objective from a low altitude.

Bursts were observed and explosions rocked the aircraft violently. The persistence and skill with which Sergeant Morgan carried out these attacks enabled the targets to be effectively bombed and ensured the safety of his aircraft.

Years later, I was a passenger in a car driving northwards over the new eight kilometre bridge connecting the south bank of the Loire estuary with St Nazaire. The drive was almost exactly like a slowed-up version of our low level attack on the oil installations. The leaf-shaped island and the oil tanks of Donges were still there and for a few moments I was back in that uncomfortable Beaufort, crouched in the nose and looking down the bombsight at the target with the yellow tracer snaking towards us.

CHAPTER TEN

Hoek van Holland

'Sir Percivale, my lord saluteth thee, and sendeth thee word that thou array thee and make thee ready, for tomorn thou must fight with the strongest champion of the world.'

SIR THOMAS MALORY. 'LE MORTE D' ARTHUR' 1469

Percival was at last pronounced fit to fly after treatment for his sinuses, which evidently consisted of some sort of unpleasant probing, and we returned to the usual routine of night flying.

On 11th October we paid what was to be a farewell visit to Brest, and the following night we called on St Nazaire for the last time. Nothing especial happened on either of these flights, and we did not realise that we were to move our base of operations. After these two trips, my leave became due, and on my return I was surprised to learn that we had been posted on detachment to Thorney Island, near Chichester in Sussex. Percival and I packed our kit and flew to this new base on 26th October.

We guessed correctly, or perhaps we were warned, that the major purpose of the detachment of most of the squadron was to place us in a position of interception if the German battle fleet passed up the Channel from Brest to their home waters. This action was some months ahead of us, however, and meanwhile we were not to be idle.

Thorney Island was a pre-war station with a well appointed mess and living quarters, and we settled in quite comfortably. Percival's wife joined him in accommodation nearby. A short distance away was the fighter station of Tangmere, with a squadron of Spitfires led by the legless Group Captain Douglas Bader. Thorney Island had seen many Beauforts before and indeed had been a training base for Coastal Command.

By now, Percival had been promoted to flight lieutenant and was acting as Officer Commanding Operations. In turn, I had been promoted to flying officer, and both of us found ourselves with an increased amount of administrative work which to some extent lessened the continual succession of our operational sorties. We flew occasionally on dive bombing practice, on air test with the new Beaufort 2, which was beginning to come off the production line at Filton, and once to St Eval and back.

Returning to Thorney Island on an occasion in the early hours of the morning, earlier than expected, I found that my room had been allocated overnight to someone taking off on operations later that morning. Not wishing to disturb this officer, I slipped down to the darkened mess and dozed off in an armchair. An hour or two later I was awakened by a roar of rage and was startled to see a very senior officer standing in front of me, although it was only about six in the morning. I leapt to my feet, and learnt that the Secretary of State for Air, Sir Archibald Sinclair, had just walked past my sleeping form, accompanied by his entourage, and that I had failed to come to attention in an officer-like manner. It was useless for me to offer explanations and I had to stand there and take an upbraiding.

Shortly afterwards, I was standing in the same mess in the early evening, drinking a half pint of bitter, when our squadron leader engineer officer said loudly to our CO:

'That boy looks terrible; he ought to be taken off operations.'

I bridled with indignation but took a close look at myself in the mirror in my room. It was true that my cheeks were somewhat sunken and my skin had a strange pallor; never more that a lightweight, my uniform now hung loosely on me. I was feeling jumpy, and the loss of almost all my friends was causing me to wonder how much longer our crew could continue without the inevitable happening. But I did not think that I was completely played out.

On 25th November we took off on a sortie that to us was remarkable. We were told that this was the first operational flight of a Beaufort 2, and we clambered eagerly into the brand new aircraft. In all respects it was the same as a Beaufort 1 except for one important difference: it would fly reliably on one engine. For months we had endured the misery of flying in an aircraft which would probably sink slowly down to the sea and cause the death of its occupant if one of its engines ceased to function. But at last the British Taurus engines were being replaced with the American Pratt and Whitney Wasp. This made little difference to the performance and indeed reduced the range by about 150 miles, but gave us a far more reliable engine and provided an important boost to the morale of the squadron.

Our patrol consisted of a flight setting off at 18.00 hours southwards to Cherbourg on the Cotentin peninsula, then skirting round the French coast towards the Channel Islands and return. We saw nothing in a rather dark night on this 'Rover' patrol but on our way back to Thorney Island we learnt that the weather had closed in. We were diverted to St Eval where we spent the night before returning the following day. The number of our Beaufort 2 was AW 248, but I am not sure if we were correctly informed about this being the first operational flight of this mark of aircraft.

The following day after our return to Thorney Island, we were briefed for an extremely dangerous sortie, a daylight 'strike' on enemy shipping. At this stage in the war, there was enormous pressure on the German railways, which had to cope with the movement of immense amounts of material to the Russian front. In consequence, the Germans were taking risks with escorted vessels carrying goods, particularly to and from the port of Rotterdam which fed the canal system of the Rhine. Our task was to look for this shipping off the Dutch coast, a new area of operations for us. We knew that losses of Beauforts on daylight shipping attacks were usually about one in three, but we were somewhat mollified to learn that an escort of Spitfires had been arranged. Our take-off time was scheduled for the following morning, 27th November. Three Beauforts were to be in vee formation, with Percival leading in our Beaufort 2, number AW 248. We were to fly to the aerodrome at Coltishall, in Norfolk, where we were to land and confer with our escorts. The remaining two Beauforts would take off an hour later and circle Coltishall and we would then all fly to the Hook of Holland and sweep north-east up to the Dutch coast.

We slept fitfully beforehand and arose at 05.00 hours on a chilly, misty night, to make our way to dispersal point and our waiting aircraft. It was still dark when we took off at 07.15, and set course for our first turning point at Petersfield, only a few miles away. From here we skirted around London, which was of course a prohibited zone, and next turned over High Wycombe, then Luton and Cambridge, before altering course slightly for Coltishall. Dawn broke en route to a very misty morning, and I was worried at first for this seemed to be thickening to fog over East Anglia. In the RAF, mist was defined as visibility below two kilometres and fog below one kilometre (3,280 feet). However, the sun had risen and the waters of the Broads gleamed through the mist, so that I was able to identify Wroxham Broad and guide Percival on the last three miles to Coltishall aerodrome, where we circled to await landing clearance.

We landed at 08.35 and conferred quickly with our escorts, who were waiting for us. Six long-range Spitfires of 152 Squadron had been ordered to escort us on the outward journey and a further six of 19 Squadron to meet us on the return flight, all flown by pilots who looked extremely young to me. I was only twenty but already an operational veteran.

We took off at 09.30 hours on our mission. Just as our wheels left the runway a flock of seagulls fluttered before us and we hit one unfortunate bird exactly on our nose, so that blood and feathers spread in all direction over my perspex windows, gradually clearing away in the slipstream.

'What was that?' Percival said sharply, as soon as the under-carriage was retracted.

'Seagull,' I replied 'Smack on our nose. No damage.'

'Just as well it wasn't the air intake,' commented Percival.

Any mishap to an engine of a Beaufort was highly dangerous, especially on take-off. If the seagull had blocked the air intake, the resulting rich petrol mixture would have caused the engine to fail, and torque from the remaining engine could have spun us, with a full bomb load, into the ground.

Five minutes later, as we circled over Coltishall, our remaining two Beauforts took up their position behind each of our wings. We were all carrying two 500 lb and four 250 lb of general purpose bombs. Above and below us ranged the six Spitfires, each armed with eight Browning .303 machine guns. For a crew such as ours, accustomed to single hit-and-run raids, this seemed a powerful formation, and we must have appeared a menacing sight as we set course to the east and the Hook of Holland. The Dutch coast was

The photograph shows one of the six 'rear cover' Spitfires of 10 Squadron which came out to meet the returning three Beauforts of 217 Squadron and the six Spitfires of 152 Squadron. This aircraft, AD185, was flown by Sgt Sokol who was probably Polish.

F/Lt Percival flying the Beaufort II, with the author in the nose of the aircraft, together with three Spitfires of 152 Squadron following. The formation of three Beauforts and six Spitfires had just left the English coast and the mist and cloud were clearing.

Time	Reqd. True Track	Distance Run	True Course	Mag. Course	OBSERVATION
1021	036		036	045	HOOK HOLLAND a/c on sweep.
1027					Ship bombed & hit 045 MASS 10
					6 MTBS Co 225
1029	293		288	297	Posn 045 MASS 10 a/c BASE
1031					Parachute in water from 020 MASS 10
1032					TAS 170 G/S 168 dist 105 Ht 500
					time 0-37 ETA 1106
1039					Bullet holes noted in starboard
					Drift 5S TMG 293.
					W/V [205 / 15] Wind
1041					1 BEAUFORT following
1043					Oil on wing tip starboard engine
1108					BASE
1125					landed

This extract from Roy Nesbit's navigation log included the following abbreviations: A/C = Alter course; MTBs = Motor Torpedo Boats (E-Boats); TAS = True air speed; G/S = Ground speed Ht = Height; ETA = Estimated time of arrival; TMG = Track made good; W/V = Wind velocity

The enemy vessel had just been bombed by two Beauforts and a third is attacking. The Dutch coast obscured by mist. The Germans on the ship had put up a spirited defence but had been overwhelmed by superior forces

only 42 minutes flying time away, and we flew at our normal operational height of 500 feet. The sun was well up above the mist and the weather cleared as we crossed the English coast at Lowestoft. There was not a cloud in the sky and the sea ahead was calm, with only a few white caps to denote breaking waves. It was going to be a beautiful day.

On this short distance, little navigation was required, or possible. I checked the wind speed and direction by frequent observation of the drift through my bombsight, and the wind lanes through my bearing compass; they were constant. Percival flew a dead straight course without the aid of the automatic pilot. We both strained our eyes ahead at the horizon. As a rough calculation, the distance of the horizon over the sea in nautical miles is the square root of one's height in feet, so that flying at 500 feet on a clear day we could see just over 22 nautical miles, or about 26 statute miles. But as we reached the Dutch coast, a slight mist, together with low strato-cumulus cloud, limited visibility. Nevertheless, we were exactly on course at the Hook of Holland and I could identify our position without difficulty as I gave Percival the alteration of course to the north-east. There was no sign of enemy ships or aircraft as the formation turned, but the Spitfires climbed and moved ahead of us.

A single dot appeared in the sea ahead. It looked like a small cargo ship, its outline becoming clearer as we approached, with two smaller ships through the mist beyond.

'About 1,500 tons, I reckon,' I said to Percival. 'Shall we attack?'

'Attacking now,' replied Percival, and rocked our wings. The other two Beauforts drew away; they did not want to follow us down to meet the explosions of our bombs fitted with eleven-second delay fuses. Little black puffs appeared in the sky. The escorts were flak ships and had opened up on us.

'1,500 tons, course 232°, painted light sea blue, one funnel,' I noted in the memoranda section my log.

This was the only occasion I did not drop the bombs during my operational flying, for it was to be a dive bombing attack. Percival climbed to starboard, then to port, turned again to port and dived. I sketched our manoeuvres in my log. The cargo ship veered vainly to starboard but there was no escape. As we dived, an orange line of tracers connected our Beaufort with the deck of the ship. For a fraction of a second, I thought that Percival was firing his front guns, and then realised that the tracer was travelling in the other direction. It was somehow different from other tracer that had been fired at us, a solid straight rod instead of a stream of bullets, and it appeared to be passing just under our nose. Percival opened up with our two forward-firing Brownings in reply, so that for a few seconds our aircraft and the ship were exchanging tracer.

Percival pulled out at about 100 feet. I saw the bombs hit the ship. There was a single courageous German seaman firing twin machine-guns at us from behind an armoured shield and I caught a glimpse of his face as Percival banked again to port and climbed. I had a clear view of the ship over my chart table as the bombs exploded and a jet of smoke and wreckage shot up from its decks. Four of our bombs had struck home and the unfortunate vessel must have nearly disintegrated, but the remaining two Beauforts dived and released their bombs on to the sinking hull. Six E-boats appeared beneath us, zig-zagging wildly and firing at us, but missing our twisting Beaufort; I noted these in my log, and worked out our return course.

'Are you all right in there, Nesbit?' Percival said sharply. There was a note of anxiety in his voice.

'Yes, of course,' I said. 'Did you see that ship go up?'

'Never mind that,' he replied. 'Come up here a moment.'

I disconnected my intercom and scrambled up to the co-pilot's seat.

'What's wrong?' I shouted.

Percival nodded to his right foot. There was a bullet hole through the outer edge of his fur lined boot, in through the front and out of the back, but no sign of blood.

'Are you all right?' I said in sudden fear. 'Shall I fly her for a bit whilst we take a look?'

'It stings a bit, but I guess I'm OK,' he replied. 'But there's something wrong with the starboard engine.'

I dived back into my compartment, from where I could see the engine more clearly. There was also something wrong with the floor of my compartment, but this was of less importance than the engine. Beneath us I quickly noted a parachute in the water; it was orange in colour and did not look like one of ours. Perhaps the Spitfires had shot down a Messerschmitt, but we had seen nothing of their battle, occupied as we were with our target. Looking over the side, I saw a line of bullet holes running down the leading edge of our starboard wing to the engine, which was emitting a thin stream of oil.

Davies' grinning face appeared behind a proffered cup of hot coffee.

'You enjoyed that didn't you, sir?' he said. 'I was watching you when the bombs exploded. Banging your fist on your table!! I've never seen you get excited before!'

I glared at him, for I did not want to lose my reputation for imperturbability.

'I took a photograph with the F 24,' he said. 'Hope it turns out OK.'

He returned cheerfully to his compartment. The F 24 was a large camera which could be operated either from a fixed position or hand held, but we could only use it in daylight since, unlike Bomber Command, we did not carry flares.

'Spitfire approaching,' reported Reeves.

The Spitfire took up station close on our starboard wing. A young face peered at us from under his helmet and goggles, and a hand waved at us.

'Thank God,' I thought. 'He can report us if we ditch.'

I jabbed my forefinger at our engine. He grinned and waved more vigorously. I picked up my Aldis lamp and flashed, 'Engine U/S.'

The Spitfire pilot executed a half roll and waved from his upside down position, demonstrating his versatility and skill in flying. He had not noticed the stream of oil, or perhaps he thought our engine always behaved that way.

'Knock it off, Nesbit,' growled Percival. 'Those guys can't read morse.'

Bored with his slow-flying charge, the Spitfire pilot zoomed off to the English coast, waggling his wings happily. A Beaufort appeared briefly behind us, but his course diverged from us. Smoke began to stream from our engine. We were alone.

'Closing down engine, feathering propeller,' said Percival.

My heart sank. It was really happening to us. A Beaufort on one engine; we would have to ditch and scramble out on to the wing to release the dinghy, and the sea would be dreadfully cold. But we were in a Beaufort 2! Percival was able to turn the propeller blades of the Wasp engine so that they were streamlined into the airflow and the effect of drag was markedly reduced. In the Beaufort 1, the propellers could not be feathered and would have continued to turn, even with the engine shut down, pumping petrol into the engine with a strong danger of fire. Our aircraft flew straight and sweetly at 500 feet, without a quiver. The sea looked as soft as silk and I returned to my trade. I checked the wind direction by wind lanes and the drift through my bombsight, and they were hardly changed. About twenty minutes to the English coast, I estimated.

When I was a small child and lived with my parents and three brothers in South Woodford, in north-east London, my father used to take us on an annual holiday to the seaside. We would board a little steam train at Liverpool Street and chug gently to the East Coast, where we stayed at a place called Gorleston-on-Sea. Two weeks was the extent of this holiday but, unlike other little boys, I never enjoyed myself very much by the seaside. I was very fair-skinned and burnt easily, did not like swimming and soon became bored playing with a bucket and spade in the sand. I used to look forward to going home to my wider selection of toys and our garden. But as we neared the English coast in our crippled Beaufort, my attitude to Gorleston-on-Sea changed completely as the little seaside town appeared dead on track, slightly on our port bow. How could I have failed to appreciate that exquisite strip of golden sand and the homely welcome of the nearby fishing port of Great Yarmouth? I glowed with pleasure as we crossed the coast with Coltishall dead ahead of us a few minutes away.

We had to circle whilst the runway was cleared, but Percival landed the Beaufort on one engine so skilfully that we could hardly feel the wheels touch the ground. The fire tender and ambulance, already alerted and racing alongside us, were not required. We were towed to dispersal point, only two hours after taking off. Percival unzipped his flying boot. There was a blue weal above his ankle, just breaking the skin. An inch to the right and the bullet would have smashed his ankle.

On 16 December 1941 217 Squadron made an attack on another convoy off the Hook of Holland, using torpedoes. Three Beauforts flew from Thorney Island to Coltishall, where they picked up an escort of three Beaufighters and twelve Spitfires.

The formation has found a convoy and the leader has just dropped his torpedo at the largest vessel, a Norwegian ship. The flak ship is firing at him.

'You'd better see the MO,' I said.

'No!' he replied. 'And don't say a word to anyone!'

I kept quiet about it. Not even Davies or Reeves knew at the time, but Percival himself must have spoken about our narrow squeak, for his slight injury is included in the official records. My guess is that Percival did not want to cause his wife any anxiety, or to present himself as a hero. He always understated our efforts at de-briefing, to the point of antagonism or contempt towards the intelligence officers.

We all climbed down from our stricken Beaufort and inspected the damage. There was the string of bullet holes along the wing and another group in the engine. But the main concentration of bullet holes was in the lower portion of the navigator's compartment. We counted thirty holes in all; several bullets must have passed just under my seat and past my legs before travelling along the fuselage, one of them nicking Percival's ankle. Davies and Reeves were completely unconcerned at our narrow escape and stood slightly to one side, chuckling together. Davies was saying loudly: 'Banging his fist on the table, he was! Our Mr Nesbit!'

We all trooped into the Operations Room for a sketchy de-briefing. The bunch of fighter pilots were laughing excitedly; they said that they had encountered some Messerschmitts and come off best.

'How large was the ship?' I was asked.

'About 1,500 tons. Maybe less,' I replied, and somebody made a note on a pad.

Our Beaufort would take weeks to repair and meanwhile we had to return to Thorney Island. Davies and Reeves crammed in with the crew of Flight Lieutenant 'Ginger' Finch, a very determined pilot whose aggression would shortly earn him a DFC for another shipping attack, during which one of our aircraft was shot down. Finch and his crew were to lose their lives only six weeks later, blasted out of the sky by the guns of the battle-cruiser *Gneisenau*, whilst flying steadily on a torpedo run towards the bows of this monstrous target.

Percival and I joined our friends Pilot Officer Seddon and Pilot Officer McGregor in the other Beaufort and we took off at 13.45

hours, cramped but happy with the outcome of our sortie, and landed just after 15.00 hours. Seddon, the pilot, a tall and rather dashing young man, was to ditch his Beaufort six miles from Malta some five months later, shot down by Messerschmitt 109s after a shipping attack. His dinghy was holed and he and the injured wireless operator stayed on the sinking aircraft and drowned. McGregor, the burly and mustachioed Canadian navigator, swam all the way to Malta although wounded by shrapnel in the back. He made it to the shore, utterly exhausted and naked, but the gunner who attempted the same feat disappeared in the sea. Later in the war, McGregor was killed in a Sunderland flying boat.

During de-briefing at Thorney Island, I looked at identification charts and revised my estimate of our ship to about 1,250 tons. It was early evening before Percival and I walked somewhat wearily into the officers' mess. The lounge bar was crowded and a group of senior oficers called to us cheerfully.

'Come over here. We're celebrating!'

Percival and I walked over and were each given a half pint of bitter. We all held our drinks by the glass, handle protruding outwards, except Percival, who held his by the handle. We all lit cigarettes, except Percival, who did not smoke.

'Cheers, sir!' I said to our CO. Percival grinned at them wolfishly.

'What are we celebrating, sir?' I asked, hoping that the squadron was about to be posted to Bermuda or South Africa, away from the war.

'Why, your success, of course!' came the reply. 'We had a signal from Coltishall. 15,000 tons! The biggest ship ever sunk by Beauforts!'

'15,000 tons?' I said, wonderingly. 'Somebody's got their decimal point in the wrong place. I told them 1,500 tons.'

Their jaws dropped and their beer glasses were lowered. Percival spluttered in his beer, then threw back his head and gave forth the succession of rasping barks that constituted his laugh. The senior officers looked at him with dislike and drifted away. That laugh probably cost Percival his DFC. Not that he would have cared.

CHAPTER ELEVEN

'Where's Digger?'

'Our life is made by the death of others,'
Leonardo da Vinci (c.1500)

During my tour of duty on 217 Squadron we lost twenty-four aircraft. Since our normal complement of crews was only about ten, the chances of survival were so slim as to be almost non-existent. On Bomber Command, it is estimated that only 30% of aircrew survived a tour of thirty sorties, but our percentage of survivors must have been far lower than even this grim statistic.

Of course, not all our losses were caused by enemy action. Without positive evidence, I believe that over half our casualties were caused by engine failure or aircrew error. In most cases we never discovered what had happened to our missing aircraft. The Germans were meticulous in announcing the names of those killed over France, but for the most part our Beauforts disappeared into the Atlantic, the Channel and the North Sea, and no more was heard of them.

On occasions, there were accidental crashes in England and these would be investigated as thoroughly as possible. One such tragedy occurred when a laden Beaufort swerved and crashed on take-off; the aircraft was tricky to fly and this was a common fault. The Beaufort caught fire but the gunner and wireless operator escaped from the mid-hatch. The pilot officer navigator was not seriously hurt at first, but his close friend the pilot was trapped in his bucket seat and was semi-conscious. The navigator wrenched open the top hatch and very bravely stayed in the flames to rescue his friend, somehow managing to drag him clear. The navigator was badly burned around the hands and arms but the pilot's injuries were much more serious and he died in hospital. In those days medical care of severe burns was improving steadily under the direction of the brilliant surgeon McIndoe at East Grinstead, but it used to be said that if an area of more than 30% of the body was burnt the result was sadly inevitable.

Whilst the navigator was recovering in hospital I went to see him, partly as my function as Squadron Navigation Officer. He was propped up in bed, bandaged but very cheerful, and after exchanging greetings asked about the condition of his pilot. I had to tell him quietly that his pilot had died. He thanked me for coming to tell him this heart-rending news, but tears began to run down his cheeks and I withdrew so as not to cause him embarrassment. What can one say to a courageous nineteen year-old officer who had tried vainly to save the life of his friend? His feat remained unrecognised and, although he recovered from his burns, he was never fit to fly again, for he developed such appalling claustrophobia that he could not bear to enter an underground train, let alone a Beaufort.

Another cause of crashes was navigational error, and I was particularly diligent in trying to prevent this type of mistake in our squadron. After my own painful experience in flying through Plymouth balloon barrage, I used to lecture all new navigators joining the squadron on this and other pitfalls. One of the common navigation errors in the RAF was setting 'red on black' on the compass. The P6 compass in front of the pilot had a rotating grid ring graduated over 360 degrees, with North designated by a red arrow head and South with a black arrow head. When the pilot had to alter course he rotated the grid ring until the required course was set against a 'lubber line' representing the fore and aft line of the aircraft and then turned the aircraft until the grid wires were lined up with the magnetic needle, with the red arrow head set against the northern crosspiece of this needle.

It was all too easy to turn the aircraft until the *black* arrow head was lined up with the northern crosspiece of the magnetic needle, so that the aircraft was flying in exactly the wrong direction, on a reciprocal course. This happened with one of our Beauforts at night when setting off from Land's End to bomb Brest with a land mine. The inexperienced pilot flew off in the opposite direction and the aircraft arrived in Eire, which is roughly the same distance away as France. And then the inexperienced navigator actually mistook Cork estuary for Brest estuary and the crew flew up this part of the

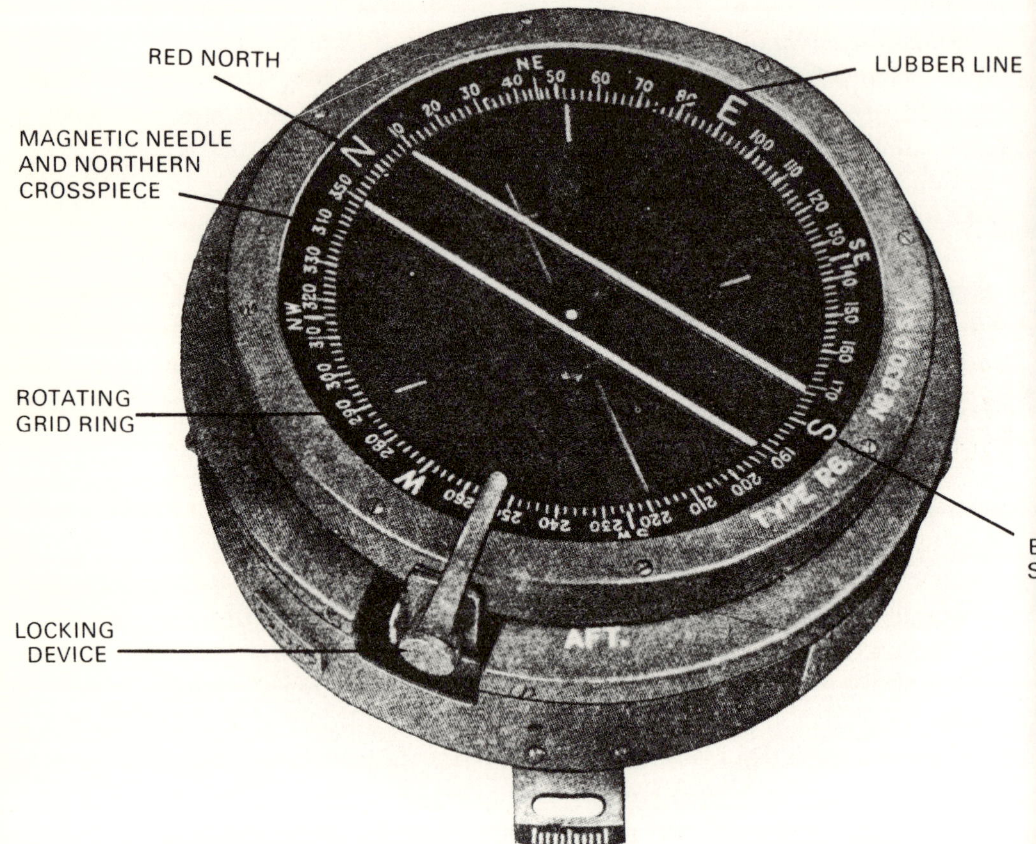

P4 and P6 compasses. These were the standard compasses for pilots in the RAF in 1941. Both were of similar construction and the P4, which was slightly bigger than the P6, was situated in front of the pilot in the Beaufort.

The magnet system and pivot were contained in a bowl filled with a mixture of alcohol and water. The north, south, east and west wires contained glass tubes filled with a luminous compound.

The outer grid ring was rotated until the course was set against the lubber line representing the fore and aft axis of the aircraft, and then secured with the clamping device.

The aircraft was then turned until the northern end of the magnet, with its cross piece, pointed to red north on the grid ring and the magnet lined up with the grid wires.

If the aircraft was inadvertently turned so that the black south on the grid ring was set against the northern end of the magnet, the pilot would be flying in the opposite direction, with potentially disastrous consequences.

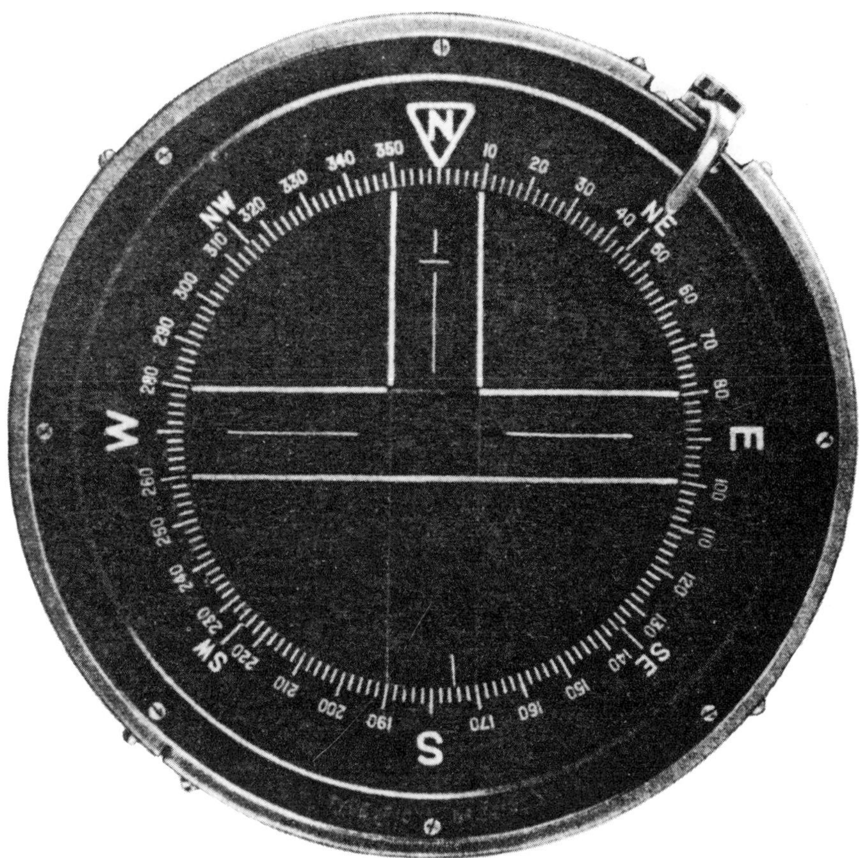

After numerous errors in setting 'red on black' on pilot's compasses, the grid ring was altered to a T shape as shown above, on a P.10 compass. This eliminated the possibility of a pilot turning the aircraft until the red south marker was lined up with the northern cross piece of the magnetic needle, when he would be flying inadvertently in the opposite direction.

This small improvement may well have saved the lives of many aircrew from 1944 onwards.

Emerald Isle and dropped their land mine, puzzled by the absence of flak; fortunately the land mine dropped in the sea and no harm was done, but this further error demonstrated the crew's incompetence. And to cap it all, the crew turned back down Cork estuary and were mystified to see dawn breaking in the west instead of the east. They called for wireless bearings and returned home in a state of shock.

Later in the war, in Southern Rhodesia, the pilot of one of our training aircraft set 'red on black' and inadvertently flew over Bechuanaland, where they ran out of fuel and had to land in the desert. The crew took out the aircraft compass and began to walk but fell in with some bushmen who murdered them for their uniforms and belongings. I later invented a simple device for correcting this fault, a tape over the southern end of the grid wires which obscured the crosspiece of the needle if wrongly set. This was introduced on many compasses and I received the princely sum of £10 for an 'inventor's award'. Later the grid wires were changed to a T shape which eradicated this common error.

Yet another form of flying error was descending blind through cloud to find one's position. Crashes occurred again and again, and still happen today, when pilots and crews adopt this suicidal practice, without other directional aids. One of our aircraft crashed in this way in southern Cornwall, killing all the crew, and I managed to find the navigation log in the wreckage, bloodstained, badly burnt but still decipherable, and took it back for analysis. The last entry was a scrawl 'attempted to get out storm', which they did by descending through cloud. When I lectured I used to plot this log on a blackboard in front of the class and then insist that every trainee look at the terrible relic in turn. I still have this log and hope that it has saved many lives.

In spite of this, the closest I ever came to death was when a pilot insisted on coming down through cloud on the approaches to Kai Tak in Hong Kong, after the Japanese capitulation, in spite of my expostulations; we missed the top of one of the islands by literally a yard or so and I could clearly see the shrubs just by the wing tip, so close that the pilot collapsed with shock against the side of the fuselage when we landed. It ended my friendship with this person. In these circumstances, if one has no approach aids, the only courses are to fly to a clearer area, or head out to sea and descend below the cloud and come in underneath it, or if this is impossible, to jump out by parachute. But never fly down through cloud to the height of the ground in that area.

I am aware of only one occasion during my tour of duty when a Beaufort landed in the North Sea or the Atlantic and the crew managed to launch the dinghy. This was when we were at Leuchars, near Edinburgh, and the pilot had to make a forced landing in the Firth of Forth. The men were all sergeants and had recently arrived on the squadron, but the pilot put the Beaufort down skilfully and three of the crew clambered into the dinghy. They were quite unharmed and yet when the rescue launch reached them, they had all died of exposure, for it was in January and the sea was icy cold. I had to lead the funeral party for one of these men and somehow we all felt resentful at what seemed to have been an unfair death. Often, of course, we had to act as 'executors' of men who failed to return, and many times I had to sort through the effects of NCOs, get rid of inappropriate items, and send their personal belongings and money back to their parents. I always included a short note of regrets, although they would already have received a letter from our CO and always there was a dignified and restrained letter of thanks from the bereaved parents. How terrible it must have been for those people to lose a son at the age of twenty or so, and yet there was never any bitterness in their replies.

Only one of our crews survived enemy fire to become prisoners of war, so far as I know. The navigator was a close friend of mine, and I was extremely relieved when the good news came to us, some weeks after they were reported missing. But on every other occasion, if the Beaufort was brought down by flak, there was no escape for the crew. At our bombing height of about 700 feet, there was no chance of baling out and no height for the pilot to try to recover an aircraft in difficulties. The Beaufort just ploughed into the ground or more usually into the sea, and it was a quick death for the occupants. Many of our aircraft must have been lost outside the harbours on mine-laying trips, and on one occasion it was reported that the Germans had exhibited the bodies of four of our airmen in public in

(Left) The quad 20mm. Th[is]
murderous flak gun had a
ceiling of about 7000' and [a]
rate of fire of about 750
explosive shells per minu[te.]

(Below) The 88mm, proba[bly]
the best known of all Ger[man]
artillery. It had an effectiv[e]
ceiling of 25,000' and a
rate of fire of 14 explosiv[e]
shells per minute.

Nantes to demonstrate their prowess at shooting down a low-flying aircraft. Perhaps this report was not true but we believed it at the time and it sent a wave of hatred through the squadron.

Of course, the loss of our friends again and again played havoc with our nerves. No doubt most young men think that they are almost immortal but we would have had to have been extremely obtuse not to realise that the scales were tipped heavily against survival. Once when returning from a flight in which almost everything seemed to have gone wrong, I was the only officer to return to the Waterbeach Hotel, for the other aircraft failed to return, and Percival lived in his cottage. I dropped off the truck and walked up the slight incline to the door, feeling sick with fear and dizzy with fatigue, at two in the morning. To my surprise a figure loomed out of the blacked-out foyer. It was my batman, or at least the batman who served three of us; we regarded him as something of a simpleton, although he did his work well enough.

'What on earth are you doing up at this time of night?' I asked.

'I heard that it was a bad night at the station, sir,' he replied, 'and I thought I'd stay up to see if you were all right. They had some beer in the NAAFI, and I thought you might like a bottle.'

He uncapped a bottle of light ale and handed it to me. I was touched and astonished but had the good sense to accept his present and not to offer to pay for it.

'It has been a very bad night,' I replied. 'But that's just what I need to revive me.'

He beamed and we wished each other goodnight. I went upstairs and sat on the edge of my bed. This small act of kindness affected me deeply but for once I could barely control the waves of fear and despair that flooded over me. I could feel myself starting to tremble and had difficulty in pulling myself together. I tried to drink the beer but it tasted metallic in my mouth and I could not finish it.

When one heard that one of the squadron has failed to return, it produced a very strange feeling, not at all as one would expect. The first impression was almost pleasurable, like a sigh of relief. I used to feel guilty about this and wonder if there was something sadistic in my nature, or a lack of compassion, and mentioned it to no one. But after the war, when there was no longer any need for pretences, I discovered that others in similar circumstances experienced much the same sensation. It would seem that it is the realisation that someone else has been taken and that you have been spared which gives rise to the feeling; perhaps it is atavistic, and maybe animals feel the same when a predator takes one of their number and they can continue grazing without fear of molestation for a while. But soon afterwards intense regrets follow and one grieves for lost comrades.

For me, the worst of these occasions was when my close friend Jock Maclean was killed. We had trained together, talked endlessly together, accompanied each other around town, and shared danger on so many occasions. Young men forge close bonds under these circumstances and his death was a bitter blow to me. Maclean and his pilot, Flight Lieutenant Halley, had been to Chivenor on a short course to learn about the new ASV (anti surface vessel) radar that was fitted to a few of our machines. A detachment from 217 Squadron was sent to Manston in Kent to experiment with this new equipment at night in the Channel. Halley failed to return and as usual we did not learn the cause. On hearing this totally unexpected news I went to see our Squadron Leader Operations and said:

'If I can get a crew together, sir, can we go and look for Halley in the Channel. Maybe they're in a dinghy.'

'You should know better than that, Nesbit,' he replied. 'Even if Maclean was your friend, we are not wasting Beauforts on daylight searches. That's a job for Air Sea Rescue.'

Of course, he was right, and I knew in my guts that the crew had gone to join the many generations of British fighting men whose graves are in the Channel.

Other than this lapse I cultivated a careful pose of indifference to danger and to casualties, and became known as a phenomenal survivor, a sort of freak. Of course, my attitude was a front for covering the quivering fear that I always felt when briefed for a sortie.

In all, I completed 49 operational flights, some quite tame, some with a strong element of danger, and some very hazardous indeed. But I must have been briefed for about 120 sorties in all, about 70 being cancelled, and also spent a great deal of time on stand-by. The effect was a constant play on my nerves and possibly I had reached

The Bofors 40mm (below) was designed and produced in Sweden. It was adapted by Britain, USA, Russia and Japan as well as Germany. The effective ceiling was about 16,000' and the rate of fire was about 100 explosive shells per minute. The German equivalent of the Bofors (left) was the 37mm flak gun.

The Beaufort carried two 500lb bombs enclosed in t bomb bay under the fusela and four 250lb underneath the wings. These are gene purpose bombs which wer less streamlined than arm piercing bombs but had a greater lateral blast.

the limit of my endurance when finally taken off operational work.

Percival was certainly not immune from fear. Once when we were walking past the small church near 217 Squadron's dispersal point, he said to me:

'Are you Church of England, Nesbit?'

'That's what it says on my identity disc,' was my reply.

'Then how about going in that church and praying?' he said.

'Praying for what?' I said, obtusely.

'Why, for our safety,' he replied.

'Not much in my line, Percy,' I said, 'but if you want to go in I'll wait here for you.'

Percival was very upset. 'Suppose I *order* you to go in,' he snapped.

I bridled up immediately. 'I suppose you *can* order me to go in, since you are a rank senior to me. But I don't think you can order me to pray!'

Percival glared at me and walked away. It was the only argument we ever had, and when I simmered down I was somewhat ashamed of myself for I ought to have gone into the church with Percival, if he felt the need for prayer, and at least supported him to that extent. My instictive feeling was that action of this sort would somehow weaken my resolve and endanger me, but perhaps I was entirely wrong.

Many of our more experienced aircrew were married and of course their wives were our guests in the station from time to time. Fortunately for me, I never had to break the news of a husband's death to any of these young wives; they left the vicinity and we heard no more of them. But it was a different matter with girl-friends. There were many Women's Auxiliary Air Force girls at St Eval, yet we saw little of them in the officer's mess, apart from the WAAF officers, who by and large were rather severe ladies. But near the Waterbeach Hotel, there was a 'Wrennery', housing a collection of extremely attractive girls in the Women's Royal Naval Service. We were not supposed to entertain these girls in our hotel-cum-mess, but authority turned a blind eye to our associations at first, and the girls all seemed able to appear in civilian dresses, probably against orders. Many liaisons developed but inevitably the girls' boyfriends were killed. Late one afternoon, I was alone in the lounge apart from

Gould, our Intelligence Officer, when a wild eyed girl burst in and screamed at me:

'Where's Digger?'

Then seeing my face, she collapsed and buried her head in the sofa, sobbing hysterically:

'Digger's dead, isn't he? I can tell he's dead!'

Digger was indeed dead, somewhere at the bottom of the Atlantic, but I did not know how to deal with such a situation and mumbled something cowardly about Digger perhaps being safe, and it was left to the fatherly Gould to escort the weeping girl back to her quarters. After this, and one or two similar episodes, the Wrens were banned from our mess. We were not altogether sorry when this happened, for like many other RAF stations we held to the strange superstition that the girl-friends of killed airmen were bad luck symbols. They were know as 'chop-girls' and we tended to avoid them, so that the poor girls suffered the twin misfortune of losing their boyfriends and being shunned by their comrades.

I have often wondered at the strange combination of circumstances which enabled Percival's crew to survive our tour of operations whilst so many other crews perished. The fates must have been with us to a large extent, for we could hardly shield ourselves against flak, apart from flying skill in weaving through the tracer and searchlights. But we never turned back from our target unless there was a valid reason for doing so and never failed to bomb accurately. There was a temptation in mine-laying to drop the 'cucumber' short of the harbour mouth, but we never succumbed to this. One of my functions was to examine the navigation logs of other navigators when they returned from an operational flight, and I sometimes suspected that the crew did not fly those last crucial few miles and so avoided the defences, but this could not be proved.

The main reason for our survival was probably our own skill, particularly Percival's ability to fly under almost any condition. For myself, I learned rapidly from my mistakes and did not repeat them, being fortunate to survive the most dangerous period of all before acquiring experience. After a while, I became almost fanatically careful, and even today when I enter a civilian air liner I note the position of the emergency exits and consider the implications of

This extraordinary photograph shows the effect of a heavy anti-aircraft shell bursting at the exact moments of its explosion. It was at high level and was fired by a German 88mm gun.

flying over different terrains. When I book a seat I always say:

'Can I sit by the emergency exit? I'm a nervous passenger.'

'Why are you nervous?' is the normal suspicious response.

'Flying in the RAF made me nervous,' I reply.

Actually the seats by the emergency exits are usually more roomy and comfortable. I always keep a mental note of our position, either by timing or by visual observation, for I can often tell where we are by looking through the window at landmarks. I enjoy these flights but retain a sense of caution rather than nervousness.

But there was an over-riding reason for survival which is difficult to explain. After a while, one becomes sensitised to danger, as if some extra ability develops. Danger can be anticipated in some mysterious way, and to some exent avoided. For instance, I seemed to know after a while whether the aircraft was ahead or behind time, or to the left or right of track, without any supporting evidence. When I saw the night fighter, it was because I could smell danger and was looking out for it. Conversely, when volunteering to fly with Morgan, I somehow sensed that it would be an effective flight. I believe that one can use fear as a sort of weapon to develop a sixth sense, provided it does not degenerate into panic. There must be an animal in each herd which can sense the predator before the others and quietly moves into the centre, leaving the imprudent to be struck down. This one will survive until it is too old and tired to run, and its special cunning is of no avail.

CHAPTER TWELVE

One-Way Ticket

'It is the savour of bread broken with comrades that makes us accept values of war.'

Saint-Exupéry (1939)

A week after our return from the Hook of Holland, I was called into the CO's office by our squadron adjutant, Flight Lieutenant Howe. After the war, for many years the only person from 217 Squadron that I ever encountered was Howe, and that was a very brief meeting outside Bush House. Almost all of the officers that I knew were killed, although there are still a few survivors from 1941 other than myself.

I marched into the CO's office and saluted.

'At ease, Nesbit,' he grunted, 'I've called you here to tell you that you're grounded.'

I gasped with astonishment and, thinking that I was being accused of 'lack of moral fibre', my first reaction was extreme anger.

'Is there something wrong with my work, sir?' I blurted.

'Nothing to do with that,' he replied, testily. 'You've done enough for the time being. Meanwhile we are disbanding your crew but you can continue as Squadron Navigation Officer and hold yourself in reserve in case we do need a navigator for operational work. You can also make yourself useful sometimes as daily duty officer. Now, what do you want to do when you leave the squadron?'

'I'd like to transfer to Bomber Command, sir,' I replied somewhat impudently. 'I'd like to fly in those new Lancasters.'

'Out of the question,' he snapped. 'I've already told you that you're off operations for a rest. You'd better become an instructor. I'll make the application. That's all.'

I saluted and went out. There is a phrase 'walking on air' which had hitherto meant no more than words to me, but now I realised its full import. An enormous weight came off my shoulders. The sensation was so pleasant that I walked all the way around the perimeter track by myself in the drizzling rain towards the officers' mess. It was a truly wonderful feeling to have survived an operational tour on Beauforts and to look forward to at least a year before going back to that flak and the enemy fighters.

I was to stay on the squadron for three months awaiting the date of my posting and was kept fairly busy with administrative work. My failing nerves, which must have been detected by my superiors, recovered surprisingly quickly. My smoking decreased and my appetite returned, but since the prospects of operational flying were limited, I developed the habit of imbibing a couple of drinks before dinner, which has persisted to this day. I also began to play squash, having been quite good at fives at school, and I still play this game.

On 7th December the Japanese attacked Pearl Harbor, and the USA entered the war. Although the Allies were retreating on almost every front, we now knew that we must win. The darkest days were over. The war might last for several more years, but we thought that the Allies must win in the end.

Davies and Reeves were posted away from the squadron. Percival was waiting to become an instructor in his native Canada. Meanwhile Percival and I flew occasionally on routine trips or on air tests. We ferried a crew up to Coltishall to bring back our Beaufort 2, now fully repaired. I navigated the CO to St Eval and back. The squadron continued to fly on operational work. January came and went, but my posting did not arrive.

On 7th February, Percival said to me: 'The CO has told me to give you some dual flying instruction. The aircraft's over there.' It was a Tiger Moth, and I had no objection to a joy ride.

'You've been solo before,' said Percival. 'Take it off.'

I took off, made a circuit, and landed smoothly.

'Not bad,' said Percival. 'Take off again.'

I took off again, climbed, banked, rolled, looped and spun. It all seemed surprisingly easy.

'You seem able to fly OK,' said Percival when he finally landed. 'I'm going to recommend your transfer to pilot.'

I burst out laughing. 'I don't want to be a pilot! I'll do my stint as an instructor and then go to Bomber Command.'

The flying bug had left me, never to return. But at least I know what had been wrong with my piloting at Prestwick; I had been too keyed up, too anxious to succeed and I had been over correcting at the controls.

The day of 12th February 1942 was a day of drama, and a humiliating episode for the RAF. I was duty officer in 217 Squadron and in the late morning the operations room began to buzz with excitement. We were told that a convoy of large enemy vessels was sailing eastwards towards Dover and that we were to attack. Our squadron at Thorney Island consisted of only four Beauforts loaded with torpedoes and three with bombs. A new CO had recently arrived but he had failed to return from his first operational flight with us, and our temporary CO was Squadron Leader George Taylor, with whom I was on good terms. Something went wrong with communications or else there was a foolish veil of secrecy, for our crews took off without being told the true nature of their target; it was the German battle fleet which had at last left Brest, and was on its way up the Channel.

When I realised the true target, after our aircraft had taken off, I could not bear to be left out. Percival was nowhere around. The only pilot available was a recently commissioned ex-sergeant whom I had helped in the squadron, but he paled and refused when I suggested that we try to get a crew together. Possibly we would not have been able to find a serviceable aircraft in any event, but I felt conscious stricken at leaving my friends to fight without trying to help.

The story of the dash of the German warships up the Channel to their home ports is so well documented that it requires little repetition. The *Scharnhorst*, *Gneisenau* and *Prinz Eugen*, escorted by seven destroyers, slipped out of Brest at 22.45 hours on 11th February. There was a radar fault in the patrolling Hudson aircraft taking off, as well as other patrol aircraft. The Germans were successful in jamming British radar stations. Adolf Galland, the German fighter ace, carefully moved his protective screen of fighters from aerodrome to aerodrome in France, Belgium and Holland to keep pace with the squadron of ships. The ships were not detected by the RAF until south of Beachy Head, near Eastbourne. From

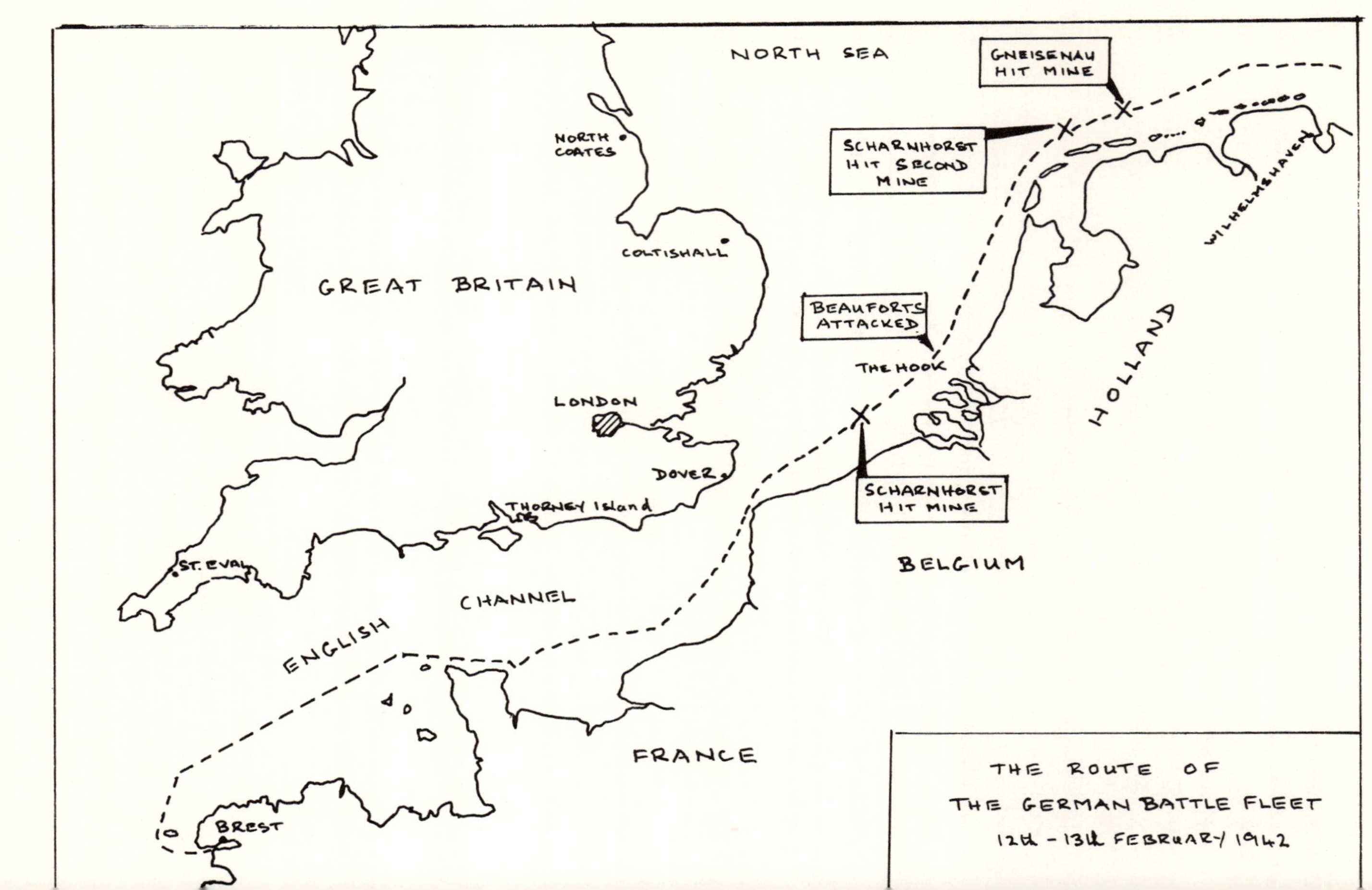

NORTH SEA
GNEISENAU HIT MINE
SCHARNHORST HIT SECOND MINE
WILHELMSHAVEN
NORTH COATES
COLTISHALL
GREAT BRITAIN
BEAUFORTS ATTACKED
THE HOOK
HOLLAND
LONDON
DOVER
SCHARNHORST HIT MINE
THORNEY Island
BELGIUM
ST. EVAL
ENGLISH CHANNEL
FRANCE
BREST
THE ROUTE OF
THE GERMAN BATTLE FLEET
12th - 13th FEBRUARY 1942

Manston in Kent six Swordfish torpedo bombers under Lieutenant Commander Esmonde attacked the ships when they were past Ramsgate, and all were shot down by flak and fighters, earning Esmonde a posthumous Victoria Cross.

Six of our Beauforts took off in the early afternoon. We lost one aircraft piloted by Flight Lieutenant Finch, shot down by flak from the *Gneisenau*, but in appalling weather conditions and against intense flak, none of our torpedoes or bombs found their target. Bomber Command launched some 250 aircraft, but scored no hits in the impossible weather and lost 15 aircraft. If the *Scharnhorst* had not struck two RAF mines, and the *Gneisenau* one mine, the victory would have been complete for the Germans. The British attacks were piecemeal and ill co-ordinated, and 217 Squadron crews were bitterly disappointed at their lack of success, for each crew had made a determined effort to attack. The depth of feeling was shown by some wag who wrote on our notice board: 'There is no truth in the rumour that the crew of the *Scharnhorst* and *Gneisenau* stopped at Brighton pier for a cup of tea during their cruise up the Channel yesterday.'

The fate of the German warships is well-known. Bomber Command scored two hits on the *Gneisenau* at Kiel, about a fortnight later, putting the vessel out of commission for the remainder of the war. The *Prinz Eugen* was torpedoed and damaged by a British submarine, en route to Trondheim, and saw no more service in the war; she was finally sunk in 1948 in the atomic bomb tests at Bikini Atoll. The *Scharnhorst* met a grim end in December 1943 when she was sunk by gunfire from British capital ships off North Cape with the loss of over 1,900 lives.

The whole episode demonstrated German superiority in planning and organisation and the ineptitude of the British command. Curiously enough however, the final advantage rested with the Allies, for the threat to the Western Approaches had been largely removed, and Bomber Command could switch from its prime target of Brest to night attacks on the heartland of Germany. But the torpedo bomber squadrons were now in bad odour in the country. The Japanese had sunk the battleships *Prince of Wales* and *Repulse* off the Malayan coast on 10th December and the question arose: If the

Japanese can sink the best of our fleet with their old bi-plane torpedo bombers surely the modern Beaufort can knock out the German battle-cruisers?

The slur on the efforts of the Beaufort squadrons caused deep resentment amongst the crews. The comparison was most unfair. The Japanese launched perhaps 100 torpedo bombers against the British battleships, which had steamed outside the radius of their possible fighter protection. We are expected to attack with maybe 10 torpedo bombers against a stronger force heavily protected by a destroyer screen and a highly efficient fighter escort in appalling weather conditions. But we were told that we must redeem the honour of the RAF and make the equivalent of kamikaze attacks against the German battle-cruisers. A suitable target was located for us. On 16th January 1942 the battleship *Tirpitz* had been moved from its berth at Kiel and had arrived in Trondheim fjord in Norway. This move was a feint to divert British naval and air forces from the area of the forthcoming dash of the ships from Brest and it succeeded. But the *Tirpitz* was somewhat more vulnerable at Trondheim than at Kiel, and 217 Squadron was sent to the very north of Scotland to deal with this target.

Percival and I flew as passengers in a Beaufort which left Thorney Island on 16th February. Our destination was a bleak aerodrome called Skitten, just north of Wick and just south of John-o-Groats in the wind-swept north-eastern tip of Scotland. Our aircraft kept developing faults, or maybe our pilot was not particularly anxious to arrive at his rendezvous. We had to put down at Driffield, Yorkshire, whilst an engine was repaired overnight, and the following day we had to make an emergency landing at Turnhouse, near Edinburgh. Rather bored, Percival and I went into Edinburgh in the evening where we finally went to a cinema. After an hour or so, we were startled to see a message flashed on the screen:

'If Flight Lieutenant Percival and Flying Officer Nesbit are in the audience, please report to the nearest police station immediately.'

Thinking that the *Tirpitz* was beckoning us, even though we were only 'first reservists' in the squadron, we hastened to the nearest police station, where the duty sergeant telephoned the RAF base at Leuchars, near St Andrews. We learnt that we had been posted as

(Left) The Beaufort is perhaps best remembered as a torpedo bomber. The torpedo which weighed about 1650lb was intended to be launched about 1000 yards from the target, and it ran at about 45 mph for 2000 yards.

(Below) In February 1942, a detachment of Beauforts of 217 Squadron was sent to Skitten, a satellite of Wick aerodrome, just south of John O'Groats, with the intention of attacking the *Tirpitz* in Trondheim Fjord.

In spite of its old-fashioned appearance, this bi-plane achieved some remarkable results for the Fleet Air Arm in World War II.
In this photograph, the Swordfish is practising torpedo dropping on British warships.

The *Tirpitz* in Trondheim Fjord.

'missing', since our pilot had failed to send a signal to 217 Squadron after he made his emergency landing.

Fortunately, the dreaded telegrams had not been sent to our relatives before our adjutant learned of our arrival at the wrong airfield.

The next day Percival and I were ferried to Skitten in an old Anson. The following week had a certain dreamlike or perhaps nightmarish quality. The *Tirpitz* was sheltering in the far end of Trondheim fjord, at a place called Fottenfjord, tucked alongside a steep rock and protected by torpedo nets and flak guns, as well as its own considerable armament. It was contemplated that our squadron would attempt to dive bomb and torpedo the battleship, and the authorities would not demur if we decided to incarcerate ourselves with our armour-piercing bombs on the deck of the 'beast' as Winston Churchill called the *Tirpitz*. Probably the attack would take place at night, but there was one over-riding snag. We had enough fuel to reach Trondheim, but it was doubtful if we would have enough fuel to get back, even if we refuelled at Sumburgh in the Shetlands, for we were likely to be extravagant with fuel over the target area with constant avoiding action. Moreover, the weather in Scotland and the Shetlands was so overcast in the depths of winter that we would be unlikely to find our home airfields without searching through the murk. However, since we were now expendable, Group Command conceived a brilliant solution; those of us who survived the attack and thought that they could not reach home should head for Sweden and jump out by parachute.

Percival and I were not to be handed these one-way tickets unless another crew went sick, an eventuality which would not have been surprising in the circumstances. Curious as it may seem, I was not adverse to making the flight, provided it was at night. I had little doubt of my ability to locate the *Tirpitz* and even bomb effectively and thought that there was a good chance of avoiding the flak and night fighters, which were likely to be less effective than at Brest. What attracted me was the prospect of a long sojourn in Sweden, and I spent much time plotting out a route over the mountains, for there is a gap beyond Trondheim fjord, and altering course over the coniferous forest belt and various towns as far as Stockholm, where there seemed to be some attractive flat pieces of land for a gentle parachute descent. If I was to be expendable, I would be so in comfort, amongst the hospitable Swedes and for as long as possible.

We were billeted in a wooden hut with a roaring fire, and for a week we sat and waited for orders. We played cards and drunk rum, the only liquor which seemed available; one day when we were off standby we walked all the way to Wick, only to find the town was 'dry' except to bona fide travellers, and evidently we did not fall into that category. Rather like a popular radio programme, we had a gramophone and about eight records, which were played incessantly. Not even my closest friends can assert that I am musical, but those tunes became so etched in my mind that even if I hear them nowadays the memory of that strange week is stirred.

Outside the hut a blizzard raged day after day, and we had to shovel a way to our neighbouring mess and other quarters. The wind howled and blew the snow into great drifts, so that the hulls of our Beauforts stood out like stranded whales above the snow. We were so far north and the clouds were so low and dark that there seemed to be only four or five hours of daylight. We could not take off, let alone fly on operations. After a week, Group decided that either we were not expendable after all, at least in this theatre of operations, or more likely that our attack was unlikely to be effectual.

The *Tirpitz* received a succession of attacks by Bomber Command and from the Fleet Air Arm operating from carriers but was unharmed. In early March 1943, the *Tirpitz* made a sortie into the Arctic and took up a new refuge at Altenfjord, near North Cape. In 1944 the vessel was hit several times by the Fleet Air Arm, and eventually was destroyed at Trömso on 12th November 1944 by Lancasters with their immense 'Tallboy' bombs of 12,000 lb, with the loss of 700 seamen.

217 Squadron flew down to Leuchars on 27th February and I was posted on 22nd March, to an instructor's course at Cranage. In May 1942 the Squadron was posted to Ceylon, to help meet the threat to India from the Japanese navy, but they were waylaid en route by the hard-pressed garrison in Malta and given the task of attacking the supply routes to Rommel's Army in North Africa. Here the squadron achieved considerable success in sinking supply ships and Italian warships, but at immense cost of lives of the crews, including almost

all the remaining airmen that I had known.

After leaving the squadron I learnt that Percival and I had been recommended for Distinguished Flying Crosses but these did not arrive and 'Mentioned in Despatches' were sent in their place. I was interested to see that these bore the signature of Sir Archibald Sinclair, Secretary of State for Air, who had been so upset when I failed to stand to attention before him. In any event, I was quite content to be a live coward rather than a dead hero and was more than happy with my survival from those dark days. After completing my instructor's course, I spent over two years in Southern Rhodesia as Station Navigation Officer at Cranbourne, near the site of what is now the airport for Salisbury. I returned home just before VE day, and after a Long Range Navigator's course, volunteered for service in a squadron operating in India, Burma, and the Far East. I was demobilized in September 1946.

I do not know what happened to Sergeant Davies and Sergeant Reeves, but I hope that they survived the war and found some happiness afterwards.

Percival was selected for posting for Canada as a staff pilot in an Air Navigation School. By 1943 he was back on operational work, in 354 Squadron, which was engaged on flying Liberators from Karachi in India over the Indian Ocean and the Bay of Bengal to counter the Japanese threat to Allied sea lanes in these areas.

On 3rd January 1944 Flight Lieutenant Percival and his crew arrived at Sigiriya, an RAF aerodrome in the heart of the jungle of Ceylon, to join 160 Squadron, which was also flying the long-range four-engined Liberators. Percival was appointed a flight commander and promoted to squadron leader.

After several operational flights, Percival took off on the dark starry night of 19th April 1944 in Liberator 'B' number BZ 864, together with his crew of eight. Just after take-off, the aircraft lost height and swung to port, its port wheel striking an AA gun site near the end of the runway, and the aircraft crashed into low trees about a hundred yards away. Within two minutes the crash tender and the ambulance had broken through the perimeter fence and the low scrub to reach the scene of the accident, followed by five men including the Commanding Officer. Two of the fliers staggered dazed but unharmed from the rear of the Liberator. The CO and his four men scrambled into the Liberator, which was beginning to burn, and pulled out four of the remaining crew, alive but unconscious; of these two died on the way to hospital and another shortly afterwards.

The fire tender was playing its extinguisher on the front of the aircraft but the fire began to burn so fiercely that rescue of the three remaining crew, including Squadron Leader Percival, were not recovered until later.

A Court of Enquiry was unable to ascertain the cause of the crash. There might have been a sudden engine failure at the critical moment of take-off, for it was difficult to service engines reliably in the tropical conditions of the jungle. Perhaps the flight engineer pulled up the lever of the flaps too quickly. Or perhaps a bird was sucked into the air intake of one of the engines.

The six officers and aircrew who died in the crash were buried with military honours in the morning of the following day, at Levermentu Cemetery, Flower Road, Colombo, Ceylon.

How can one describe the young men of 217 Squadron in 1941? Since I was one of them, it is perhaps not my place to comment. But, a few years ago, I was a guest at a dinner party when those present began to discuss the mentality of RAF aircrew engaged on bombing attacks in the last war, not knowing that I had been one of this number. The most vocal of these conversationalists was a woman doctor, who expressed her opinion in no uncertain terms:

'Of course, they must all have been psychopaths.'

This profound opinion met with general agreement from the remainder of the guest, most of whom doubtless considered themselves to be intellectuals. I spoilt the party, and indeed broke it up, by remarking that the level of intelligence in RAF aircrew seemed to me considerably higher than that of the company I saw around me that evening.

In contrast to the opinion of the 'intellectuals', I have also met several young people who are fascinated by World War 2 but who tell me that in reading about the conflict they have come to the conclusion that Britain must have been populated by a race of

A four-engined Liberator bomber taking off from an airfield in Southern India. It was in such an aircraft that the author's pilot Squadron Leader John Frederick Percival was killed on take-off at Sigiriya aerodrome in central Ceylon on the night of 19 April 1944.

supermen whose deeds they could never hope to emulate. For my part, I doubt if there is much difference in the qualities of courage or the degree of sanity possessed by either generation. It seems to me that young men expose themselves to danger in wartime for a variety of reasons, such as patriotism coupled with territorial fighting instincts, the desire to impress their friends and relatives, or simply just curiosity about themselves – a wish to find out how they will respond to conditions of extreme duress and hazard.

Having experienced a year of great danger, I am certain that I cannot be classed as a hero but I still do not know if I am a coward. It seems to have been external circumstances which brought out cowardice or what might be regarded as bravery in me. For instance, as an eleven year old school boy I was once chased home by a gang of 'yobbos', as we called young thugs in those days, who evidently thought that my school blazer designated me as belonging to a stratum of society more effete that their own. I was terrified, but as the leading yobbo chased me through our front gate, a great rage overcame me and I swung round and struck my pursuer such a blow with my left fist that he was knocked flat on his back. He picked himself up, blood trickling between the fingers that he clasped to his injured nose, and rushed away howling, followed by his startled cronies, his departure being accelerated by a well-aimed kick that I directed at his backside. It was extremely satisfactory for me to be able to walk unmolested throught the neighbourhood after that episode, but I could not understand what had prompted my sudden outburst of courage, especially since I had been at the point of reaching complete safety; it must have been a sort of territorial defence mechanism coupled with a dread of my parents and brothers witnessing my abject terror.

In the same way, I would attempt to fight ferociously in my own defence in the RAF, especially when near home, but my courage dwindled away when flying far from base. Not within my capabilities was the incredible bravery of Guy Gibson, flying his Lancaster round the Möhne dam in an attempt to draw the flak away from other aircraft running in to drop their 'bouncing bombs' accurately. My policy was to be the first aircraft to arrive under the radar screen into an unsuspecting target at night, in case somebody had stirred up the defences like a wasps' nest, or to be the last in formation when attacking a target by day, in the hope that the flak gunners would direct most of their fire against the preceding aircraft. Of course, if one was ordered to attack in a certain order one was compelled to obey, but given a free choice I would always try to think out and opt for the safest plan of action.

In many ways my attitude of extreme caution was probably the most effective, not only for my survival, but to add to the sum total of blows that were struck against the enemy. Most crews must have had similar thoughts, and I rarely met any airmen who were longing to get to grips with the enemy or to risk their necks unnecessarily. Only once did such a crew arrive in the squadron, in the middle of 1941, and although the pilot and navigator were both unassuming and likable young officers, they seemed to go out of their way to court danger, like knights in former times challenging champions in an opposing army to leave the ranks and fight them. Ordered on patrol work, they would break off and intrude into enemy ports in daylight to drop bombs on shipping, or they would hunt out and attack enemy aircraft that outclassed them. They survived only a few operational flights before they lost their Beaufort and their own lives, earning posthumous DFC's. There was no doubt as to their courage, but perhaps they would have done more damage to the enemy if they had been less impetuous and survived longer.

No fliers that I met in 217 Squadron appeared to want to kill the enemy for the sake of killing, although doubtless some aircrew from Occupied Europe had conceived such a hatred of the Germans that this was their prime motive. Although most of the fliers in 217 Squadron would not have admitted weakness at the time, they were probably almost as scared as myself. If they had been psychopaths they would not have had the intelligence to do their job, and in any event fliers without fear would probably have killed themselves accidently during training. Possibly bravery consists of conquering one's own fears, and most young men of 217 Squadron managed to do this. They were mostly highly trained amateurs-turned-professionals who were fully aware of the risks they were running, but were patriotic enough and possessed sufficient qualities of responsibility and comradeship to offer their lives in a fight against a brutal and totalitarian regime.

I was not the best or bravest of these young men but I am proud to have been in their company.

INDEX.